PHASOR METHODS OF AC CIRCUIT ANALYSIS

- DESIGNED USING MATLAB OBJECT ORIENTED PROGRAMMING

SHORT INDEX

PREFACE

APPROACH

DC and AC circuit analysis is the beginning fundamental course in the Engineering/Technology programs. Many textbooks use calculators as the engine for calculations which require long and complex calculations. In the long calculations, students often loose attention while their immediate concerns become not to commit mathematical mistakes. As a consequence, they do not pay enough attention on the concepts and tools. Furthermore, the conventional method of ac circuit analysis use phasors and phasor based methods. Some Instructors use MATLAB as the calculation engine but it is primarily designed to work with complex numbers and methods. This book presents a Phasor Tool Box which contains most functions that would be needed to do phasor calculations and enable students to visualize in the phasor diagrams. The tool box is designed in MATLAB and requires students to have minimal scripting background, keeping in mind that these tools will be used by the beginner students in Electrical engineering/Technology programs. This tool box employs object oriented MATLAB programming methods but are transparent to users. Using these tools is as easy as using algebra for adding, subtraction, multiplication, division of phasors, and additionally visualize these operations in the complex plane. It is hoped that introduction of the phasor methods would help in fortifying the conceptual learning.

This book is not intended to be a textbook but it can be used as the co-book. This book is aimed at the students who are just beginning circuit analysis in the undergraduate program of engineering/technology and have either working knowledge of MATLAB programming or have worked on the tutorial in Appendix A and D prior to starting on Chapter 1. Code of all Phasor Tool Box functions are given in Appendix E and can also be downloaded (free download) from https://professorjaiagrawal.weebly.com/phasor-methods-in-ac-circuit-analysisfirst-course-in-digital-control.html.

.

OTHER BOOKS BY PROF. AGRAWAL

First Course in Digital Control: using MATLAB/SIMULINK

ISBN-13: 978-1546724964
ISBN-10: 1546724966

Marketed by Creatspace.com/Amazon.com
ISBN-13: 978-1511651271
ISBN-10: 151165127X

ACKNOWLEDGEMENTS

This book is dedicated to my students who have given me strength after strength. I will like to thank my colleague Prof. Omer Farook who encouraged me to write this book.

CHAPTER 1

INTRODUCTION TO PHASOR TOOL BOX

LEARNING OBJECTIVES

- To define vector, arrays and matrices of phasor, polar or complex variables and quantities.
- To perform mathematical operations on the phasors/polars
- To plot the phasor, polar or complex vector on the complex plane
- To plot the phasor, polar or complex vector in the time-domain and frequency domain
- To convert from one form to other forms

CHAPTER INDEX

AC circuit analysis involves manipulations of sinusoidal waveforms, complex and polar vectors. Visualization uses time domain waveforms and vector diagrams on the complex plane. Engineering education programs often uses MATLAB tools which are basically designed for complex and matrix manipulations. Polar variables and quantities are often converted to complex Cartesian forms for any processing, rendering complicated and long effort. This book describes the direct polar methods that utilizes Object Oriented Programming and functions in MATLAB for ac circuit analysis. The Object Oriented programming uses objects to represent the polar quantities and variables. Visualizations employ

I will like to make a point about the concepts of circuit and network at the beginning of the book. An electrical network is an interconnection of electrical elements such as resistors, inductors, capacitors, transmission lines, voltage sources, current sources, and switches. An electrical circuit is a network that has a closed loop, giving a return path for the current. A network is a connection of two or more components, and may not necessarily be a circuit.

We will not make this differentiation and refer both as circuits in this book.

1.1 VECTOR

A vector is a directed arrow from point a (x1, y1) to another point b (x2, y2) on the complex plane. Plot of a vector on the complex plane is obtained in the MATLAB from the code,

MATLAB_Ex_1.1
Draw a vector from point (1, 2) to point (4, -4) on the complex plane.

```
%PTB2_Ex_6.m
%A vector
%arrow is a standard MATLAB function.
axis([-6, 6, -6, 6])          %Specify the area of the complex plane
arrow([1, 2], [4, -4])        %a vector from point (1, 2) to point (4, -4) on the complex plane
grid
axis('square')
%arrow is MATLAB function, which can be downloaded from the Matlab central website
```

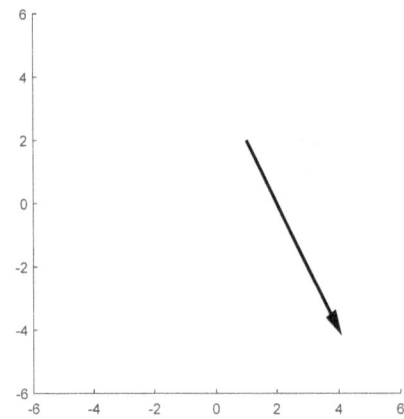

MATLAB_Ex_1.2

Draw two vectors on the complex plane: A - from point (1, 2) to point (4, -4) and B - from point (1, 2) to point (5, 4) on a single complex plane.

```
%PTB2_Ex_7.m
%Two vectors
clf
axis([-6, 6, -6, 6])          %Specify the area of the complex plane
axis('square')
arrow([1, 2], [4, -4])        %a vector A from point (1, 2) to point (4, -4) on the complex plane
hold                          %hold the plot size and scale
arrow([1, 2], [5, 4])         %a vector B from point (1, 2) to point (4, -4) on the complex plane
grid
hold                          %release the hold
```

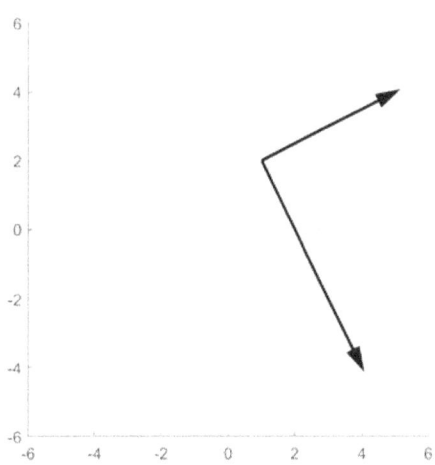

1.2 COMPLEX, POLAR VECTORS AND PHASOR

1.2.1   COMPLEX AND POLAR VECTOR

MATLAB is basically designed for operations on matrices and complex numbers. The phasor tool box has a function **phplot()** to plot a complex quantity on the complex plane as a directed arrow in which the x-component is the real magnitude and the y-component is the imaginary magnitude. The arrow originates at point (0, 0) on the complex plane. We will call it the directed arrow from origin a phasor, and define it as **class** in this book.

MATLAB_Ex_1.2.1

Draw a complex quantity C=3+j4 on the complex plane. Find the magnitude and angle in degrees of the vector.

```
%PTB2_Ex_8.m
Clf, clear
C= 3+j*4;                     %C is a complex quantity
```

```
phplot(C, 'cx')                    %Plot of the complex quantity C on the complex-plane,
C_mag=abs(C)                       % magnitude of the vector
C_angle=rad2deg(angle(C))          %angle in degrees, with respect to the horizontal line, and counterclockwise
% the rad2deg() is a standard function in MATLAB
```

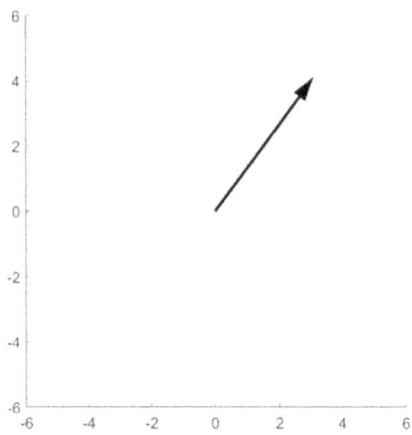

AC circuit analysis involves sinusoidal waveforms and variables. Consider an ac voltage signal is given by:

In sinusoidal waveform format, $\quad v(t) = V_m \sin(\omega t + \theta)$

In the exponential format, $\qquad\quad v(t) = V_m e^{j(\omega t + \theta)}$

where V_m is the peak voltage, $\omega = 2\pi f$, f the frequency in Hz and θ is the phase angle in radians. If the frequency f is constant, the signal amplitude varies in a sinusoidal manner and the phase angle varies from 0 to 2π radians over and over again as time proceeds. Since ω is constant, we can freeze it or assume it to be zero without loss of any information, the resulting expression is called the vector V which rotates at an angular frequency ω:

Exponential Vector: $\qquad V = V_m\, e^{j\theta} \qquad\qquad$ where θ is denoted in radians

$\qquad\qquad\qquad\qquad\qquad$ in MATLAB: $\qquad\qquad$ V1=Vm*exp(j*th)

This vector can be visualized as a vector of length V_m rotating on a circle from 0 degree to 360 degrees in the counterclockwise direction as time proceeds. The vector angle with respect to the horizontal axis represents the varying time. This vector is often written by engineers in a simpler format, named the polar:

Polar Vector: $\qquad\qquad V = V_m \angle\theta \qquad\qquad$ where V_m is the peak voltage and θ the angle in degrees.

The polar vector is also expressed in Cartesian components in the complex format,

Complex Vector: $\qquad\quad V = V_m \cos\theta + j\, V_m\, \sin\theta$

This polar vector is used to model voltage, current, impedance, admittance and power. For voltage and current, electrical engineers popularly use rms (root mean square) value in place of the peak value. The polar vector with rms value is named the Phasor.

Phasor $V = V_{rms} \angle \theta$

The phasor does not include any information of the frequency of the signal, which also means that the phasor is independent the signal frequency. The following figure shows various voltage phasors and their corresponding sinusoidal signal waveforms with appropriate phase angles:

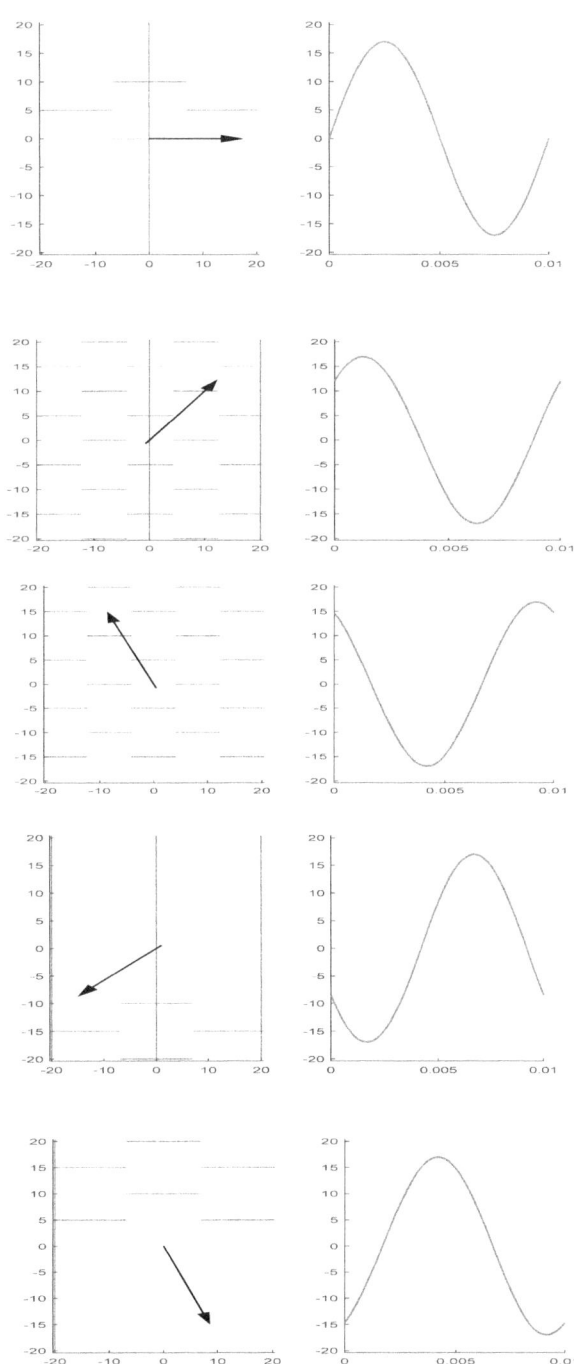

Fig. 1.1 Phasors in the complex plane and their corresponding sinusoidal signal waveforms with appropriate phase

angles in the time domain.

1.2.2   PHASOR TOOL BOX FUNCTIONS

The Phasor Tool Box has a class variable $\boldsymbol{phasor}(A, th)$ to model voltage, current, impedance, admittance and the apparent power vectors. The following list provides the Phasor Tool Box codes for different vector types:

Variable	Vector type	Math expression	Phasor Tool Box code
Voltage V	phasor	$V = V_{rms} \angle \theta$	phasor(Vrms, th)
Current I	phasor	$I = I_{rms} \angle \theta$	phasor(Irms, th)
Impedance Z	polar	$Z = Z_m \angle \theta$	phasor(Zm, th)
Admittance Y	polar	$Y = Y_m \angle \theta$	phasor(Ym, th)
Power S	polar	$S = S_m \angle \theta$	phasor(Sm, th)

PHASOR TOOL BOX FUNCTION: *phasor(rm, rp, type)*
A class to specify a phasor or a polar object. The object has two properties, *rm* and *rp,* which are respectively the magnitude and the phase angle in degrees with respect to a horizontal line in the counterclockwise direction on the complex plane.

phasor(rm, rp):   Defines a phasor or a polar object. The property *rm* In phasor objects is rms (root mean square) value. The property *rm* In polar objects is peak value and rp is the phase angle in degrees.
phasor(x, y, 'x2ph'):   Defines a phasor or a polar object from complex vector. The property x is the real value, y is the imaginary value. The property Magnitude of the produced object is the rms (root mean square) value.
phasor(x, y, 'x2po')   Defines a phasor or a polar object from complex vector. The property *r* is the real value, *i* is the imaginary value. The property Magnitude of the produced object is the absolute value of the complex vector. Phasors can be concatenated into an array or matrix of phasors.
phasor(r, 0):   Defines a dc object. The property *r* is the magnitude.

PHASOR TOOL BOX FUNCTION: *phplot(X, type)*
Plots X as an arrow on the complex plane. The object X can be a phasor or a complex vector.

phplot(X)   *X is either a phasor or a polar matrix.* In a phasor X(rm, rp), the arrow originates at the point (0, 0) and points at a counterclockwise angle of rp degrees with respect to the horizontal line on the complex plane. The rm is length of the arrow.
phplot(X, 'cx')   *X is a complex matrix.* In a complex vector X(=a+j b), the arrow originates at the point (0, 0) and terminates at (a, b) on the complex plane. Arrow length is the absolute magnitude of the complex vector.

PHASOR TOOL BOX FUNCTION: *triplot(A)*
Plots A as a triangle of arrows on the complex plane. The object A can be a phasor, polar or a complex vector or matrix.

triplot(A)   Plot of phasor/polar matrix A in the complex plane

PHASOR TOOL BOX FUNCTION: **VT=add_graph(V1, V2, V3, …)**

*P*hasor V1 is plotted on the complex plane from the origin (0, 0), then V2 is added to the end of V1. Phasor V3 is added to the end of V2, and so on. Finally, the phasor VT, the algebraic sum of all V1, V2, …, is drawn from the origin to the end of the last vector added. No limit on the number of vectors for addition.

MATLAB_Ex_1.2.2

Make and plot the following phasor and polar objects and complex vector:
a)   Phasor object for voltage V with a magnitude of 12 Vrms and phase 45 degrees.
b)   Phasor object for current I with a peak value of 5 Amp and phase 30 degrees (lagging).
c)   Polar object for impedance Z with a peak magnitude of 10 ohms with 60 degree phase angle.
d)   Polar object for admittance Y with a peak magnitude of 0.1 Siemens with -30 degree phase angle.
e)   Polar object for apparent power S in a circuit of 1 kVA and phase 30 degrees (lagging).
f)   Phasor from complex voltage X with a real part of 12 V and imaginary part 8 V.

```
%PTB2_Ex_9.mlx
clf, clear
%%Given information
Vrms=12;        %peak magnitude of a sinusoidal ac
voltage
th=45;          %phase of the ac voltage in degrees
%%%solution
V=phasor(Vrms, th)      % V is phasor
V =
  phasor with properties:
    Mag: 12
    phase: 45
%alternately V1=phasor(12, 45);
phplot(V)       %Plot of the phasor V on the
complex-plane
Current plot held
Current plot released


clf
triplot(V) %triangle plot of phasor V
```

```
clear, clf
Im=5;                   %peak magnitude of the current phasor
Irms=Im/sqrt(2);
th=30;                  %phase of the Current phasor in degrees
I=phasor(Irms, th)      % I is phasor
I =
  phasor with properties:
    Mag: 3.5355
    phase: 30
%alternately I=phasor(5/sqrt(2), 30);
triplot(I)  %triangle plot of phasor I
```

```
clf
Zm=10;                  % magnitude of impedance in ohms
th=60;                  %phase of the impedance in degrees
Z=phasor(Zm, th)        % Z is polar
Z =
  phasor with properties:
    Mag: 10
    phase: 60
%alternately Z=phasor(10, 60);
triplot(Z)  %triangle plot of polar Z
```

```
clf
Ym=0.1;        %peak magnitude of an
admittance
th=-30;          %phase of the admittance
in degrees
Y=phasor(Ym, th)          % Y is polar
Y =
  phasor with properties:
     Mag: 0.1000
     phase: -30
%%alternately Y =phasor(0.1, 30);
triplot(Y) %triangle plot of polar Y
```

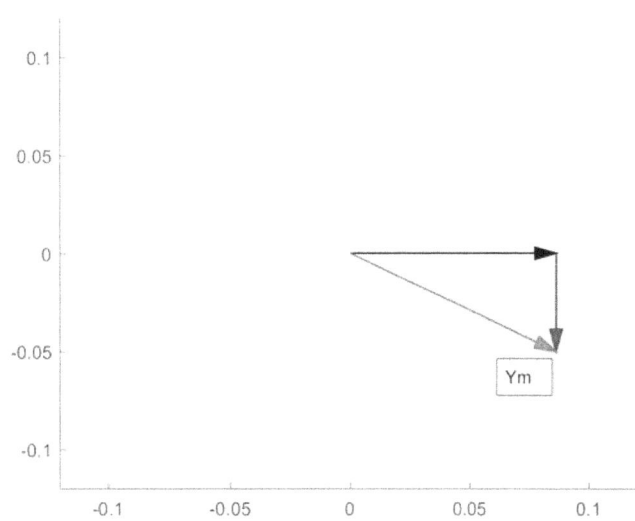

```
clf
Sm=1000;       %peak magnitude of an apparent power
th=30;            %phase of the apparent power in degrees
S=phasor(Sm, th)          % S is polar
S =
  phasor with properties:
     Mag: 1000
     phase: 30
%%alternately S =phasor(1e3, 30);
triplot(S) %triangle plot of polar S
```

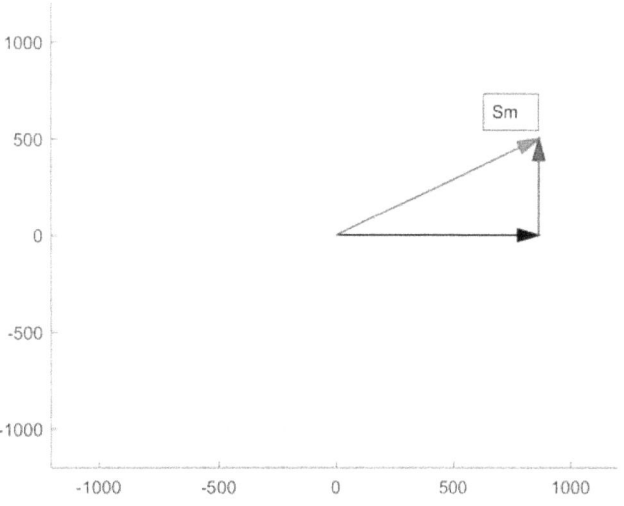

```
clf
X=12+j*8;          % Complex voltage
A=phasor(12, 8, 'x2ph') %complex
voltage is defined as phasor
A =
  phasor with properties:

     Mag: 10.1980
   phase: 33.6901
phplot(A)          %Plot of A on the
complex-plane
Current plot held
Current plot released
clf
triplot(A)   %triangular plot of A on the
complex-plane
```

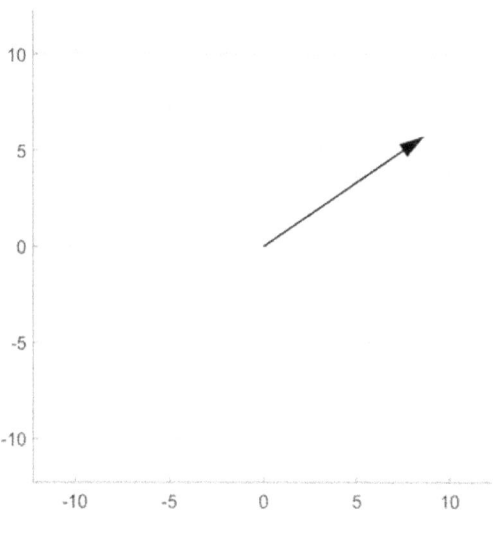

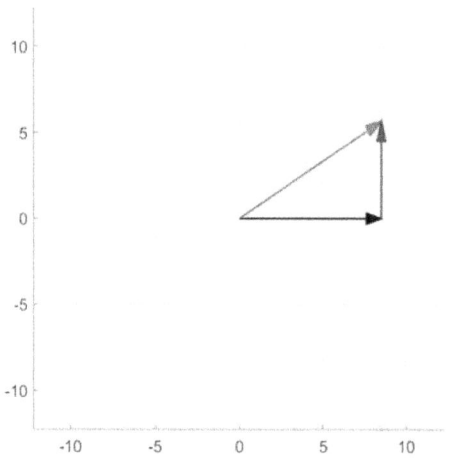

Following example shows plotting of array and matrix of Voltage/current phasors.

MATLAB_Ex_1.2.3
Draw an array of three voltage phasors V1 = (12, 45), V2=(5, -30) and V3=(10, 150) on a single complex plane. Define
V1, V2 and V3 as the elements of a phasor array V.

```
%PTB2_Ex_10.mlx
clf, clear
f=100;
V(1)=phasor(12, 45);
V(2)=phasor(5, -30);
V(3)=phasor(10, 120);
phplot(V)                    %plot the array of phasors
```

Current plot held
Current plot released

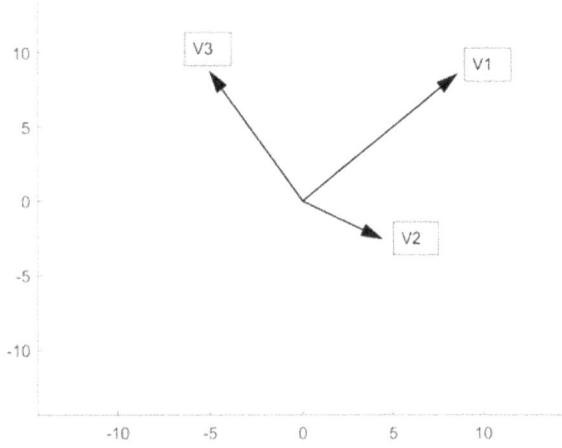

Following example shows the plotting of matrix of voltage/current phasors.

MATLAB_Ex_1.2.4
Draw a matrix of 2x2 voltage phasors and add them graphically.

$$V = \begin{bmatrix} (12,45) & (5,-30) \\ (10,120) & (8,-120) \end{bmatrix}$$

```
%PTB2_Ex_10A.mlx
clf, clear
f=100;
V(1, 1)=phasor(12, 45);
V(1, 2)=phasor(5, -30);
V(2, 1)=phasor(10, 120);
V(2, 2)=phasor(8, -120);
phplot(V)        %plot the array of phasors
Current plot held
Current plot released
```

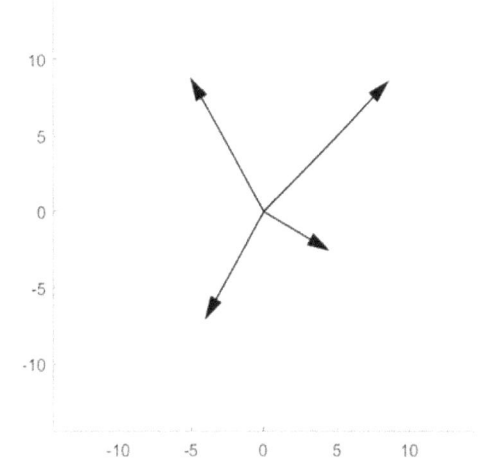

VT=add_graph(V(1,1), V(1, 2), V(2, 1), V(2, 2))   %VT (in red) is the algebraic addition of all phasors V(i, j).

VT =
  phasor with properties:

    Mag: 8.6090
    phase: 63.6925

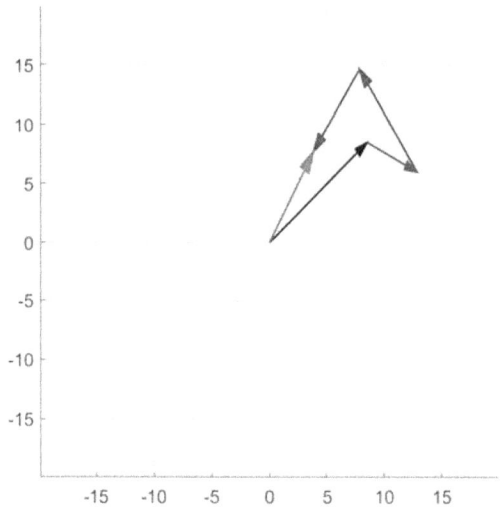

MATLAB_Ex_1.2.5
Draw three-phase voltage phasors of 12 Vrms ∠45. Add all three phase phasors graphically.

The three phasors when added graphically, produce a zero voltage phasor.

```
%PTB2_Ex_10A.mlx
clf, clear
V(1)=phasor(12, 45);
V(2)=phasor(12, 45-120);
V(3)=phasor(12, 45-240);
phplot(V)      %plot the array of phasors
Current plot held
Current plot released
```

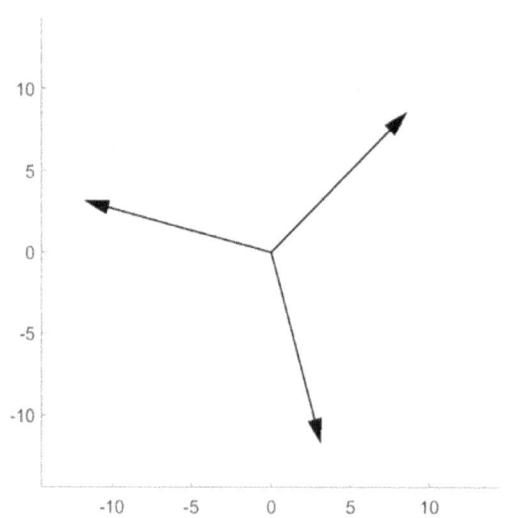

add_graph(V(1), V(2), V(3))
ans =
  phasor with properties:

    Mag: 2.5121e-15
   phase: 225

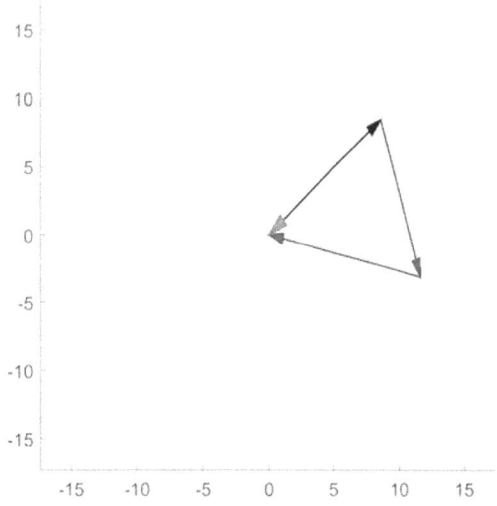

1.3   FORM CONVERSIONS

1.3.1   COMPLEX ↔ PHASOR

PHASOR TOOL BOX FUNCTION: **P=x2*ph(X, type)***
Converts a complex matrix **X** to phasor/polar matrix **P**. The input argument **X** is a complex matrix.
P=x2ph(obj)   Convert a complex matrix to phasor matrix.
P=x2ph(obj, 'po')   Convert a complex matrix to polar matrix.

PHASOR TOOL BOX FUNCTION: **X=ph2x*(P, type)***
Converts a phasor/polar matrix **P** to complex matrix **X**. The input argument **P** is a phasor/polar matrix.
x=ph2x(obj)   Convert a phasor matrix to complex matrix.
x=ph2x(obj, 'po')   Convert a polar matrix to complex form.

MATLAB_Ex_1.3.1:   Convert a) Complex vector C=3+j 4 to a phasor b) Phasor 10 ∠45 to a complex value.

%PTB2_Ex_1AA.m
%path(path,'c:\Users\jpagrawa\Documents\Aschool\Books_JP\PhasorToolBook\PTB4')
%a) Complex to phasor
C1=3+j*4;
C1p=x2ph(C1)
C1p =
  phasor with properties:

    Mag: 3.5355
   phase: 53.1301

```
%b) Phasor to complex
C2p=phasor(10, 45);
C2c=ph2x(C2p)
C2c =
  10.0000 +10.0000i
```

1.3.2   COMPLEX ↔POLAR (IMPEDANCE (Z), ADMITTANCE (Y) OR APPARENT POWER (S))

MATLAB_Ex_1.3.2   Convert c) Complex vector Sx=3+j 4 to a polar d) Polar Z=10 ∠45 to a complex value.

```
%PTB2_Ex_1AA.m
%path(path,'c:\Users\jpagrawa\Documents\Aschool\Books_JP\PhasorToolBook\PTB4')
%c) Polar to complex
Sx=3+j*4;
Sp=x2ph(Sx, 'po')
Sp =
  phasor with properties:

    Mag: 5
  phase: 53.1301
%d) Polar to complex
Zp=phasor(10, 45);
Zx=ph2x(Zp, 'po')
Zx =
  7.0711 + 7.0711i
```

1.3.3   COMPLEX ARRAY ↔ PHASOR ARRAY

Converting complex array to voltage /current phasor array and vice versa:

MATLAB_Ex_1.3.3   Convert a) Voltage phasor array V =[6 ∠ − 30 , 10 ∠45] to a complex array X.
                          b) Complex array X=[3+j*4; 4-j*5] to a phasor array V.

```
%PTB2_Ex_1B.mlx
%a) Voltage Phasor array to complex array
V1=phasor(6, -30);  V2=phasor(10, 45);
V=[V1, V2];
X=ph2x([V1, V2])
X =
  7.3485 - 4.2426i  10.0000 +10.0000i
%b) Complex array to Phasor array
X1=3+j*4; X2=4-j*5;
X=[X1, X2];
V=x2ph(X);
V(1), V(2)
```

ans =
  phasor with properties:

    Mag: 3.5355
   phase: 53.1301
ans =
  phasor with properties:

    Mag: 4.5277
   phase: -51.3402

1.3.4   COMPLEX ARRAY ↔ POLAR ARRAY

Converting complex array to ZYS phasor array and vice versa:

MATLAB_Ex_1.3.4   Convert a) Impedance polar array Zp =[6 ∠ − 30 , 10 ∠45] to a complex array Zx.
                                   b) Complex array X=[3+j*4; 4-j*5] to a Power array S.

```
%PTB2_Ex_1D.mlx
%Polar array to complex array
Zp=[phasor(6, -30),  phasor(10, 45)];
Zx=ph2x(Zp, 'po')
Zx =
  5.1962 - 3.0000i   7.0711 + 7.0711i
%Complex array to polar array
X=[3+j*4, 4-j*5];
S=x2ph(X, 'po');
S(1), S(2)
ans =
  phasor with properties:

    Mag: 5
   phase: 53.1301
ans =
  phasor with properties:

    Mag: 6.4031
   phase: -51.3402
```

1.3.5 COMPLEX MATRIX ↔ PHASOR MATRIX

Converting complex matrix to voltage /current phasor matrix and vice versa:

MATLAB_Ex_1.3.5   Convert

a) Voltage phasor matrix to a complex matrix X.

$$V = \begin{bmatrix} 6\angle -30 & 10\angle 45 \\ 5\angle 60 & 8\angle 135 \end{bmatrix}$$

b) Complex matrix to a phasor matrix V.

$$X = \begin{bmatrix} 3+j\,4 & 4-j\,5 \\ 3-j\,4 & 5-j\,5 \end{bmatrix}$$

```
%PTB2_Ex_1F.m
%a) Voltage Phasor matrix to complex matrix
V11=phasor(6, -30);  V12=phasor(10, 45);  V21=phasor(5, 60);  V22=phasor(8, 135);
V=[V11,    V12
   V21,    V22];
X=ph2x(V)

%b) Complex matrix to Phasor matrix
X11=3+j*4; X12=4-j*5;  X21=3-j*4; X22=5-j*5;
X=[X11,  X12
   X21,   X22];
V=x2ph(X)
V(1, 1),  V(1, 2)
```

```
X =
  7.3485 - 4.2426i  10.0000 +10.0000i
  3.5355 + 6.1237i  -8.0000 + 8.0000i

ans =
  phasor with properties:
    Mag: 3.5355
   phase: 53.1301
ans =
  phasor with properties:
    Mag: 4.5277
   phase: -51.3402
```

1.3.6   GET PROPERTIES OF PHASORS IN A MATRIX

Obtain the properties of a phasor matrix.

MATLAB_Ex_1.3.5
Find the properties all phasors in the following voltage matrix:

$$V = \begin{bmatrix} 6\angle -30 & 10\angle 45 \\ 5\angle 60 & 8\angle 135 \end{bmatrix}$$

```
%PTB2_Ex_1G.m
%a) Voltage Phasor matrix t
V11=phasor(6, -30);  V12=phasor(10, 45);  V21=phasor(5, 60);  V22=phasor(8, 135);
V=[V11,    V12
   V21,    V22];
[B, C] = prop(V)     %B is the matrix containing magnitudes and C matrix contains the phase of all elemental phasors

B =
   6   10
   5    8
C =
  -30   45
   60  135
```

1.4   PHASOR MATH

This section presents various mathematical operations involving phasors.

1.4.1   UNARY PHASOR MATH

This section shows the operators that work on the phasor variables.

1.4.1.1 AMPLIFYING/MULTIPLYING PHASOR BY A CONSTANT

A coefficient is often needed in phasor calculations, therefore we define it as a phasor with zero phase. The Phasor Tool Box has a function coeff() for this purpose.

MATLAB_Ex_1.4.1.1   Multiply a phasor X=(12, 45) by 2 and plot.

```
%PTB2_Ex_11.m
% amplifying a phasor by a constant
clf
X=phasor(12, 45);
C=coeff(2);        %define a coefficient
```

```
%Alternately C=phasor(2, 0);
phplot([X, C * X])        %Plot on the complex plane
```

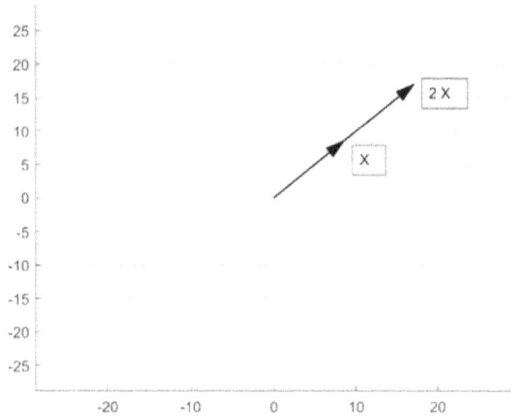

1.4.1.2 ROTATING PHASOR

A phasor may need rotation on the complex plane. The Phasor Tool Box has a function rotate() for this purpose.

MATLAB_Ex_1.4.1.2   Rotate a phasor X=(12, 45) by 30 degrees in the clockwise direction.

```
%PTB2_Ex_11.m
clf
X=phasor(12, 45);
p=rotate(30);          %rotate phasor X counterclockwise by 30 deg
Xp=X*p
phplot([X, X*p])       %Plot on the complex plane
%}
```

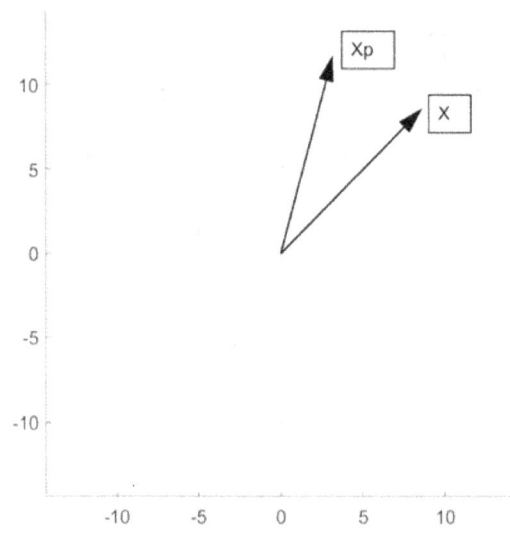

1.4.1.3   NEGATION

Negation of a phasor is to change its direction by 180 degrees. Phasor Tool Book has an operator '-' for this operation.

MATLAB_Ex_1.4.1.3   Plot a phasor X=(12, 45) and its negated phasor -X.

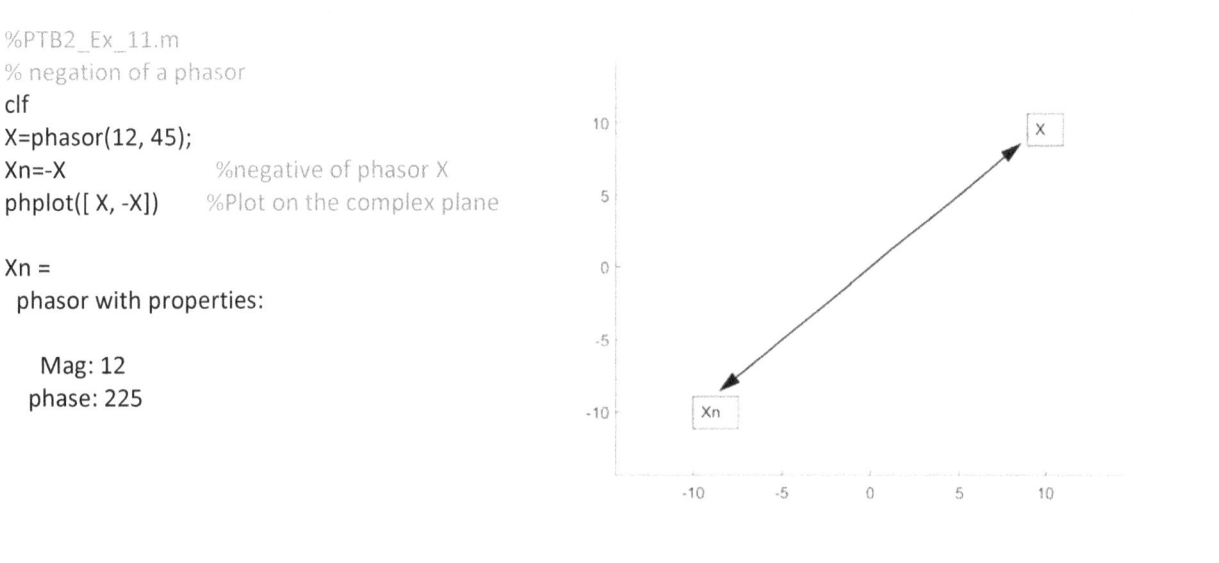

```
%PTB2_Ex_11.m
% negation of a phasor
clf
X=phasor(12, 45);
Xn=-X               %negative of phasor X
phplot([ X, -X])        %Plot on the complex plane


Xn =
 phasor with properties:

   Mag: 12
  phase: 225
```

1.4.1.4   CONJUGATE OF A PHASOR

Conjugate of a phasor X is another phasor with the same magnitude but with negative phase. Phasor Tool Book has a function conj() for this conversion.

MATLAB_Ex_1.4.1.4:   Plot a phasor X=(12, 45) and its conjugate on the same complex plane.

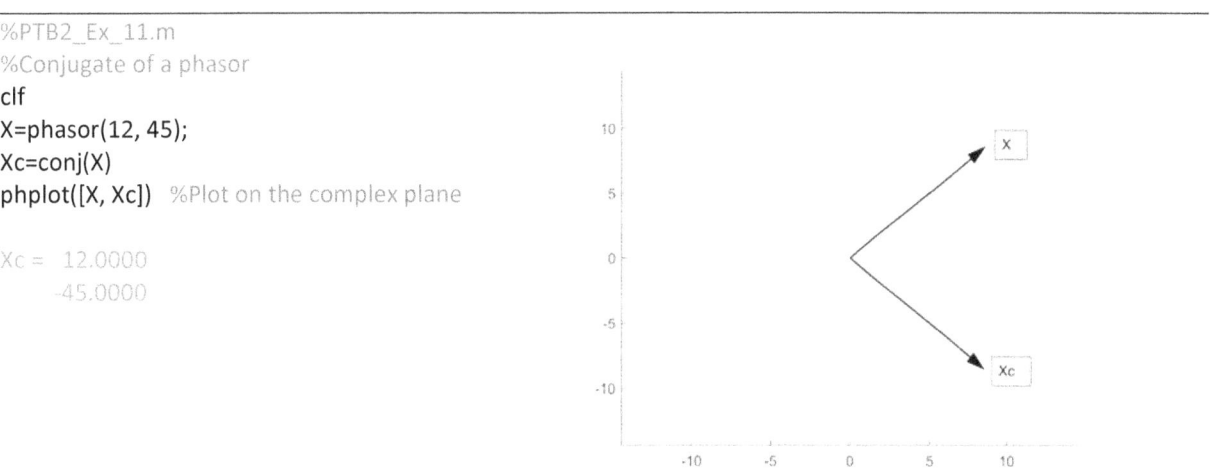

```
%PTB2_Ex_11.m
%Conjugate of a phasor
clf
X=phasor(12, 45);
Xc=conj(X)
phplot([X, Xc])   %Plot on the complex plane

Xc =   12.0000
     -45.0000
```

1.4.2   BINARY PHASOR MATH

1.4.2.1   ADDITION/SUBTRACTION OF PHASORS

Phasor Tool Book uses '+' and '-' operators for addition and subtraction respectively. It is binary operation but can be repeated on one line code of the MATLAB for Nary operation.

MATLAB_Ex_1.4.2.1:   Calculate and plot the addition and subtraction of two phasors X1= 12 Vrms ∠45 and X2= 5 Vrms ∠120.

```
%Addition and subtraction of two
phasors
clf
X1=phasor(12, 45);    X2=phasor(5,
120);
X1+X2
phplot([X1, X2, X1+X2])    %Plot on
the complex plane
X1-X2
%phplot([X1, X2, X1-X2])    %Plot on
the complex plane

ans =
  phasor with properties:

    Mag: 14.1442
    phase: 64.9656

ans =
  phasor with properties:

    Mag: 11.7449
    phase: 20.7190
```

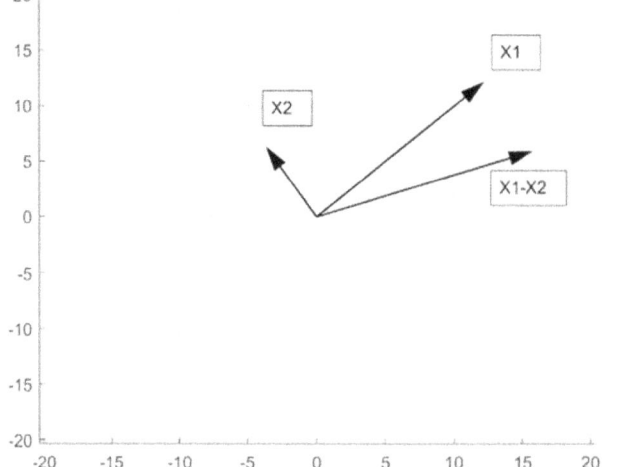

1.4.2.2 MULTIPLICATION OF PHASORS

Phasor Tool BoX uses '*' operator for multiplying two phasors. It is binary operation but can be repeated on one line code of the MATLAB for Nary operation.

MATLAB_Ex_1.4.2.2:   Multiply three phasors V1= 120 V ∠60, V2= 90 V ∠30 and V3= 50 V ∠-45.

```
%PTB2_Ex_11.m
%multiplication of two or more phasors
clf
V1=phasor(120, 60);
V2=phasor(90, 30);
V3=phasor(50, -45);
Vmul=V1 * V2 * V3

Vmul =
  phasor with properties:
    Mag: 540000
    phase: 45
```

1.4.2.3 DIVISION OF PHASORS

Phasor Tool Book uses '/' operator for division in two phasors. A/B divides phasor A by phasor B; the magnitude of A is divided by the magnitude of and the phase of B is subtracted from the phase of A. It is binary operation but can be repeated on one line code of the MATLAB for Nary operation.

MATLAB_Ex_1.4.2.3:   Divide phasor V1= 120 V ∠60 by V2= 90 V ∠30.

```
%PTB2_Ex_11.m
%Divison of two phasors
clf
V1=phasor(120, 60);
V2=phasor(90, 30);
Vdiv=V1 / V2

Vdiv =
  phasor with properties:
    Mag: 1.3333
    phase: 30
```

1.4.2.4 INVERSION OF A PHASOR

Inversion of a phasor X produces the phasor 1/X, the magnitude is 1 /|X| and the phase is negative.

MATLAB_Ex_1.4.2.4:   Plot a phasor X=(12, 45) and its inverted phasor, amplify the magnitude of the inverted phasor by 100 times in order to obtain a reasonable phasor diagram on a single complex plane.

```
%PTB2_Ex_11.m
%inversion of a phasor
clf
X=phasor(12, 45);
C=coeff(100);
Xi= C / X
phplot([X, Xi])      %Plot on the complex plane
%the amplitude of Xi is amplified 100 times for a good plot.
Xi =   0.0833
     -45.0000
```

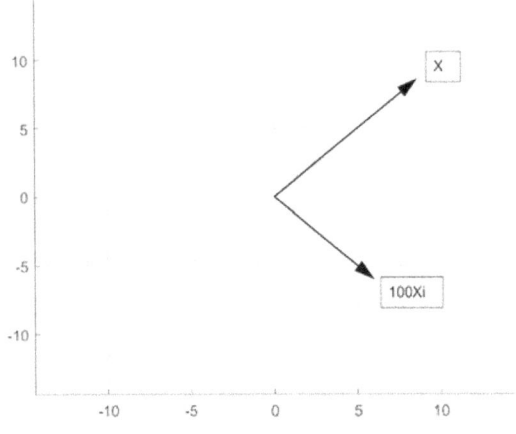

Note: 1) Parentheses () can be used in phasor math like a regular math operation for prioritization of operations on a MATLAB line code.

2) (Phasor expression) is evaluated but the result is not a phasor unless assigned. Therefore, in order to access properties, we should first assign E=(phasor expression) and then access its properties as E.Mag or E.phase. That is, (phasor expression).Mag is not permitted.

MATLAB_Ex_1.4.2.5:   Evaluate and plot a phasor expression:

$$V_n = \frac{120 \angle 60 \; (90 \angle 30 \; - 50 \angle -45 \;)}{90 \angle 30 + 50 \angle 45 + 100 \angle 0}$$

```
%PTB2_Ex_11.m
%A complex phasor math expression
clf
V1=phasor(120, 60);
V2=phasor(90, 30);
V3=phasor(50, -45);
V4=phasor(100, 0);
Vn=V1 *(V2-V3) / (V2 + conj(V3) + V4)
phplot([V1, V2, V3, V4, Vn])
```

Vn =
  phasor with
properties:

  Mag: 47.8791
  phase: 101.4343

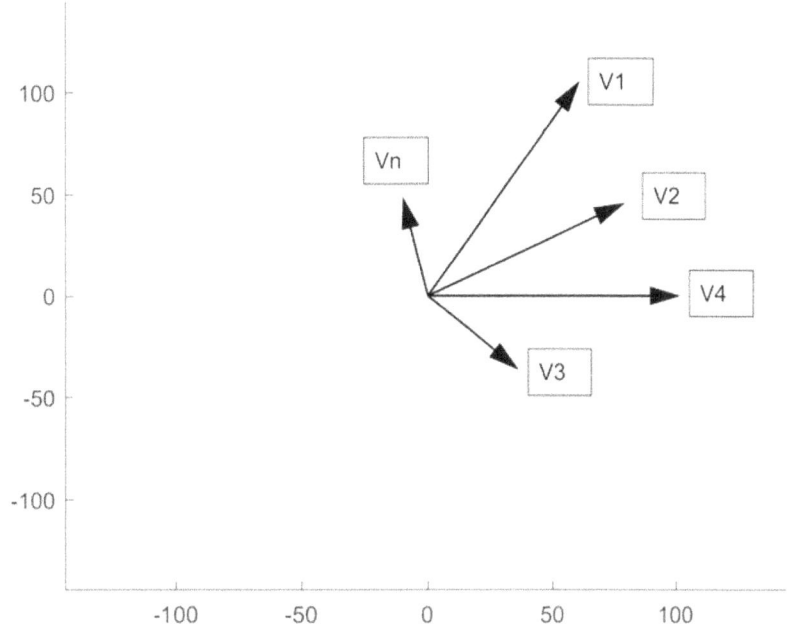

1.4.2.5   SOLUTION OF SIMULTANEOUS EQUATIONS IN PHASORS

The solution of simultaneous equations AX=B in phasor is obtained from

   $X=A\backslash B$

The Phasor Tool Box has method in the 'left divide operator' to do this job.

MATLAB_Ex_1.4.2.5:
A circuit has following equation: **Z I = E**
where the impedance matrix **Z** and the excitation vector **E** are given as

$$Z = \begin{bmatrix} 1\angle 30 & -5\angle 45 \\ -5\angle 45 & 1.5\angle 145 \end{bmatrix} \qquad E = \begin{bmatrix} 20\angle 60 \\ -10\angle 0 \end{bmatrix}$$

Find the solution $I = \begin{bmatrix} I(1) \\ I(2) \end{bmatrix}$

```
%PTB2_Ex_40C.m
 Z11=phasor(1., 30);   Z12=phasor(5, 45);
 Z21=Z12;              Z22=phasor(1.5, 145);
E1=phasor(20, 60);     E2=phasor(10, 0);
Z=[Z11,  -Z12,
  -Z21,   Z22 ];
E=[E1; -E2];                 % E is a column vector
```

```
%solution for loop currents
I=Z\E ;                         %Solution of Z I=E equations
I(1), I(2)

ans =
  phasor with properties:
    Mag: 2.2382
    phase: -49.0374
ans =
  phasor with properties:
    Mag: 2.7783
    phase: -155.8986
```

Next is an example of calculating the real power delivered to an ac circuit.

MATLAB_Ex_1.4.2.6
Draw the phasors V and I on the complex plane and find the real power delivered to the following circuit:

$$I=(7, -45)$$

+

$$V=(12, 30)$$

-

Fig. 1.1

```
%PTB2_Ex_14B.mlx
clf
f=60;  T=1/f;
V=phasor(12, 30);
I=phasor(7, -45);
S=V * conj(I)    %apparent power
S =
  phasor with properties:

    Mag: 84
    phase: 75
phplot([V, I])
Power=S.Mag * cosd(S.phase)
Power = 21.7408
```

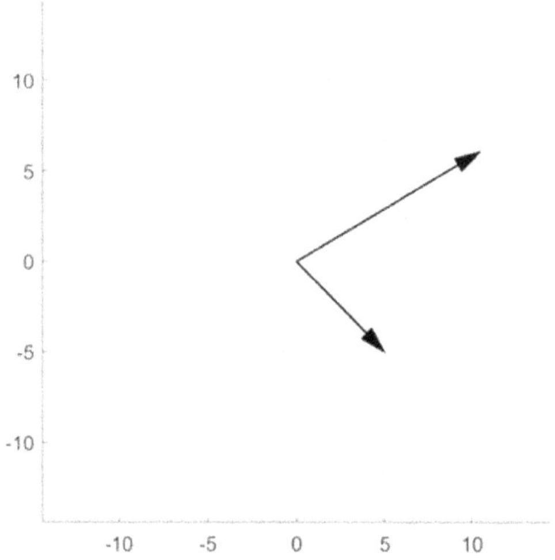

1.5   TIME DOMAIN PLOT OF PHASORS

The time domain plot of phasor requires the specification of frequency of the ac signal. The Phasor Tool box function to plot the time domain signal is phplot_signal().

PHASOR TOOL BOX FUNCTION: **phplot_signal***(V, f, t1, t2, type)*

[y, t]=phplot_signal(V, f, t1, t2)   plots the sinusoidal waveforms of matrix of phasors **V** over a range of time between t1 and t2. The y and t are respectively the amplitude and the time matrices of the time domain signal of the phasor V y is the amplitudes of the matrix of phasors in V.
[y, t]=phplot_signal(Vx, f, t1, t2 , 'cx')   plots the sinusoidal waveforms of matrix of complex vectors **Vx** over a range of time between t1 and t2.

V or **Vx** may be a single phasor, a phasor array or a phasor matrix.

MATLAB_Ex_1.5.1
Draw a phasor V1 with a magnitude of 120 V peak and phase 45 degrees in the complex plane and its time domain signal over two periods if the frequency of the signal is 100 Hz.

```
%PTB2_Ex_2.m
%Sept13_ECET15200_4
clf
f=100; %frequency in Hz
V=phasor(120, 45);
subplot(121)
phplot(V)
subplot(122)
phplot_signal(V, f, 0, 2/f);
```

Another example of ac circuit analysis.

MATLAB_Ex_1.5.2
Calculate and plot voltage VL, current IL across an impedance ZL= (3, 90) ohms, on the complex plane and the time domain signal. You may like to amplify the current phasor by 5 in order to obtain a good graph on a single complex plane and time domain. The signal frequency is 60 Hz.

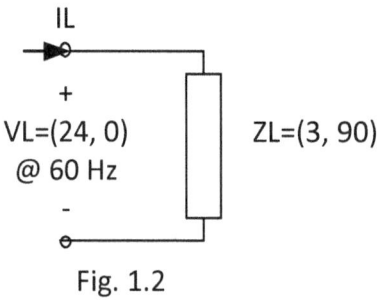

IL

+

VL=(24, 0)
@ 60 Hz

ZL=(3, 90)

-

Fig. 1.2

```
%PTB2_Ex_4.m
%ECET15200_Sept14_1
clf
f=60;  T=1/f;
VL=phasor(24, 0);
ZL=phasor(3, 90);
IL=VL / ZL      %VL/ZL
subplot(211)
phplot([VL, IL]);
subplot(212)
phplot_signal([VL, IL], f, 0, 2*T);


IL =
  phasor with properties:

    Mag: 8
    phase: -90
```

An example of time-domain plot of phasor matrix.

MATLAB_Ex_1.5.3   Draw the time domain signal of a phasor matrix V over two periods if the frequency of the signals is 100 Hz.

```
%PTB2_Ex_10A.m
Clf, clear
f=100;
V(1, 1)=phasor(12, 45);
V(1, 2)=phasor(5, -30);
V(2, 1)=phasor(10, 120);
V(2, 2)=phasor(8, -120);
phplot(V)                  %plot the array of phasors
clf
phplot_signal(V, f, 0, 1/f);
```

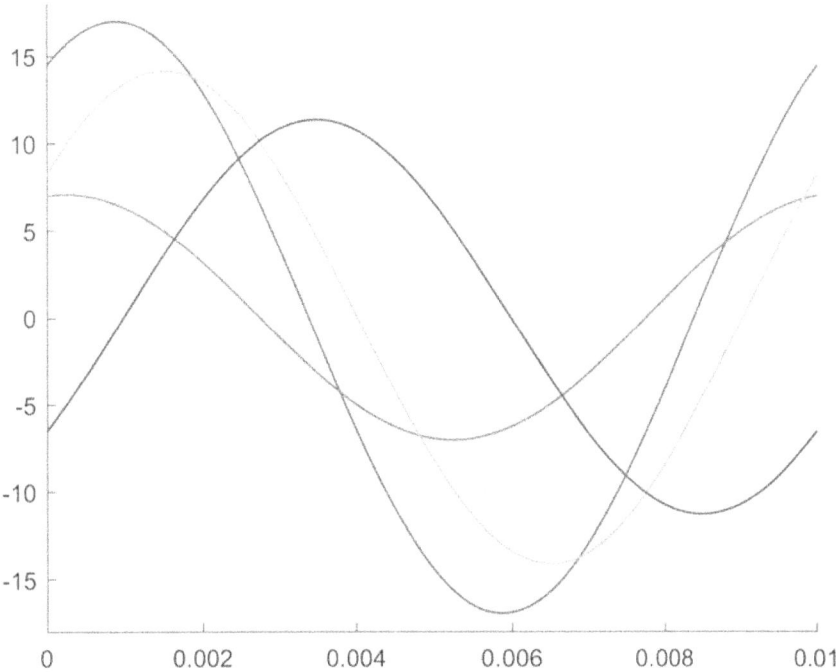

1.6   FREQUENCY DOMAIN PLOT OF PHASOR

Frequency spectrum of phasors require the specification of frequency of the ac signals.

MATLAB_Ex_1.6.1   Get the frequency domain plot of current phasor in the following circuit. Assume the frequency range of 100 Hz to 20 kHz.

```
%PTB2_Ex_3.m
clf
f=100:10:2000;                    %range of frequencies
R=10; L=100e-3; C=1e-6;
Z=R+ j*2*pi*f*L- j ./(2*pi*f*C);  %impedance
subplot(211)
plot(f, abs(Z), 'linewidth', 1 )  %impedance magnitude vs. frequency
grid
Vp=phasor(120, 30);               %voltage in phasor form
V=ph2x(Vp);                       %voltage in the complex form
I=V ./ Z;                         %current in the complex form
subplot(212)
plot(f, abs(I), 'linewidth', 1 )  %current magnitude vs. frequency
grid
```

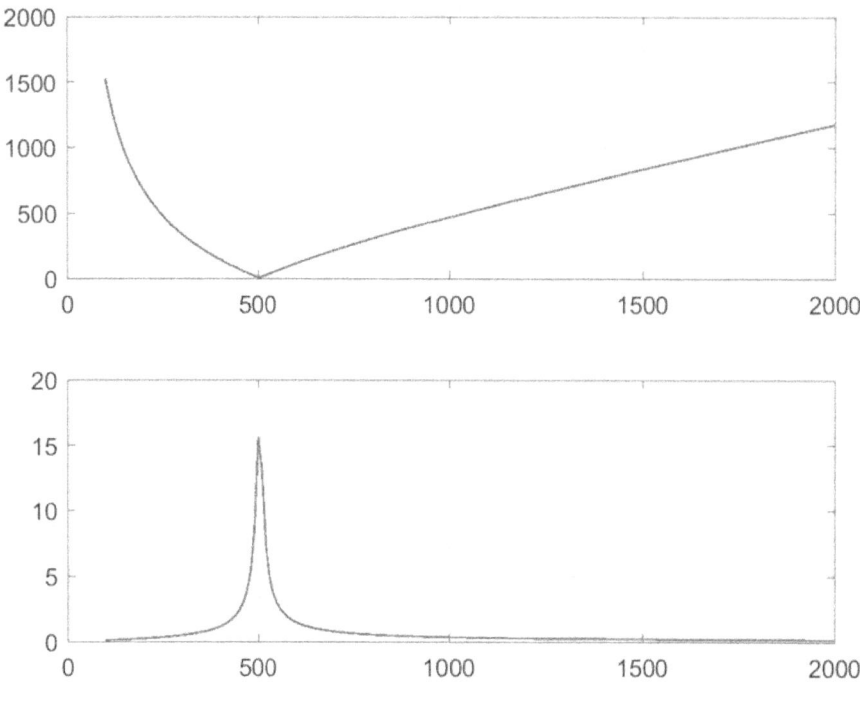

_____

1.7   LIST OF PHASOR TOOL BOOK OPERATORS AND FUNCTIONS

The Phasor Tool Book contains operators and functions that will perform most of the tasks encountered in the ac and dc circuit analysis:

Conversion functions: such as the complex to polar and phasor and back.
Math operations on phasor variables: add, subtract, multiply and divide.
Complex plane plots: of polar and phasor quantities
Time domain plots of phasors
Series-Parallel circuit: determining voltage and current in all components and finding input impedances.
Thevenin and Norton equivalent circuits
Power calculations: in single and three phase circuits

Full list of Phasor Tool Box functions is as the following:

	PHASOR TOOL BOOK MATH OPERATOR	Description
1	+	Add two phasors, the binary operator
2	-	Rotating a phasor by 180 degrees, the unary operator, also Subtract one phasor from another, the binary operator
3	*	Multiplying two phasors

4	/	Dividing one phasor by another phasor
5	\	Used in X=A\B for solving AX=B simultaneous equations.

	MATLAB function	Description
1	phasor(rm, rp, type)	A class object with Magnitude rm and phase rp. If 'type' is 'x2ph' or 'x2po' the rm and rp are the real and imaginary parts.
2	coeff(c)	Multiply a phasor amplitude by c.
3	rotate(d)	Rotate a phasor by d degrees.
4	x2ph()	Conversion from Complex to Phasor /Polar form, x is a single complex value, an array or a matrix of complex values
5	ph2x()	Conversion from Phasor /Polar to Complex form, P is a single phasor, array or a matrix of phasors.
6	ph2pu()	Phasor to per unit converter
7	pu2ph()	Per unit to phasor converter
8	conj(P)	Conjugate of a phasor, P is a single phasor/polar quantity or an array of phasor/polar quantities
9	triplot(V)	Draw the triangle of a phasor/polarctor (V, I, Z, Y or S), the real component and the imaginary component in the complex form as the base and the height and the phasor magnitude as the hypotenuse of the triangle.
10	add_graph(V1, V2, …)	Phasor V1 is plotted on the complex plane from the origin (0, 0), then V2 is added to the end of V1. Phasor V3 is added to the end of V2, and so on. Finally, the phasor VT, the algebraic sum of all V1, V2, …, is drawn from the origin to the end of the last vector added. No limit on the number of vectors for addition. VT is in red color.
11	phplot(V)	Plot of phasor/polar vector, arrays and matrix on the complex plane.
12	phplot_signal (V, f, t1, t2, type)	Plot of phasor/polar/complex vector **V** of frequency **f** in time domain, over a range of time between t1 and t2, V may be a single phasor, an array or a matrix of phasors.
13	PWR (V, I, type)	Calculate the apparent power VA, VAR, Real Power and power factor, 'lagging or leading', V and I are voltage and current phasors/complex vectors into an element or a subcircuit
14	line2phase(EAB, EBC, ECA)	Line voltage to phase voltage converter in 3-phase ac circuits
15	phase2line(Ean, Ebn, Ecn)	Phase voltage to Line voltage converter in 3-phase ac circuit
16	PWR_3phase(Vph, Iph)	Calculate apparent power S, reactive power Q and the real power P, the power factor Fp in a 3-phase circuit, Vph and Iph are arrays of the phase to neutral phasors in all 3 phases.
17	PWR_3line(VLL, IL)	Calculate apparent power S, reactive power Q and the real power P, the power factor Fp in a 3-phase circuit, VLL and IL are arrays of Line to line phasor voltages and line current phasors in all 3 phases.
18	parallelZ (Z, type)	Parallel combination of two or more impedances, Z is an array of impedances in phasor/complex form.
19	inputZ (Z, V)	Input impedance in complex form across **a-a'** terminals of the voltage **V** in a circuit that has an impedance matrix **Z.**

20	Thevenin(Z, E, V)	Find the Thevenin equivalent circuit across **a-a'** terminals in a circuit, Z is the impedance matrix with terminals **a-a'** shorted, V is source vector with **a-a'** shorted.
21	delta2wye(ZA, ZB, ZC, type)	Delta to Wye circuit impedance converter in complex form
22	wye2delta(Z1, Z2, Z3, type)	Wye to Delta circuit impedance converter in complex form

CHAPTER 2

R-L-C CIRCUITS

LEARNING OBJECTIVES

- Work with Impedance and admittance in polar form and plot the impedance diagrams on the complex plane.
- Combining impedances in series and parallel circuits.
- Input and output impedance in circuits
- Voltage and current phasors in R-L-C circuits, phasor diagrams and time-domain waveforms.
- Voltage and current divider rules
- Y-delta circuits

CHAPTER INDEX

AC circuit analysis requires working with R-L-C circuits in which the components may be specified in all mixed formats: complex, polar and phasor and the same is also true with the voltage and current sources. In this chapter we will work mostly in the phasor/polar format and use conversions whenever necessary.

2.1   IMPEDANCE AND ADMITTANCE

2.1.1   IMPEDANCE

The impedance of an electrical component is represented in a complex form or a polar form:

Impedance is a complex number, with the same units as resistance, for which the SI unit is the ohm (Ω). Its symbol is usually Z, and it may be represented by writing its magnitude and phase in the form $|Z| \angle \theta$. The real part is called the resistance, which is independent of frequency. The imaginary part is called reactance, which is dependent on frequency. It can also be represented by a polar vector on the complex plane, where, the length of the vector gives the magnitude and the angle with reference to the horizontal axis in the anticlockwise direction.

The Phasor Tool Box function to specify is *phasor(Pm, Ph)* in which Pm is the magnitude of the impedance.

MATLAB_Ex_2.1.1
Find the magnitude, phase angle in degrees of an impedance which is the series combination of 1000 ohms and 4 Henrys inductance at 100 Hz. Plot impedance Z as a polar vector and the impedance triangle on the complex plane.

```
%PTB2_Ex_110.mlx
clf
R=1000;  L=4;  f=100;
Z=phasor(R, 2*pi*f*L, 'x2po')
Z =
  phasor with properties:
    Mag: 2.7049e+03
   phase: 68.3030
phplot(Z)
Current plot held
Current plot released
```

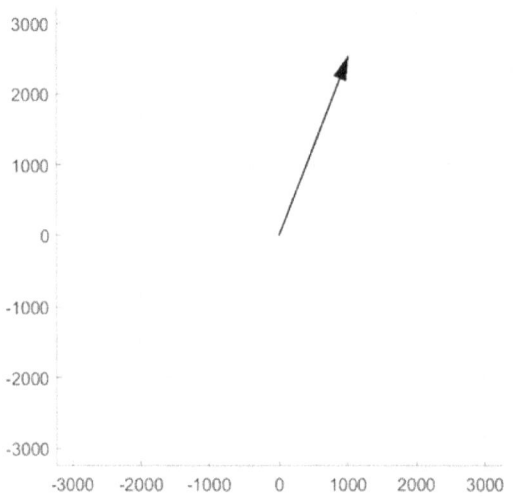

```
clf
triplot(Z)
```

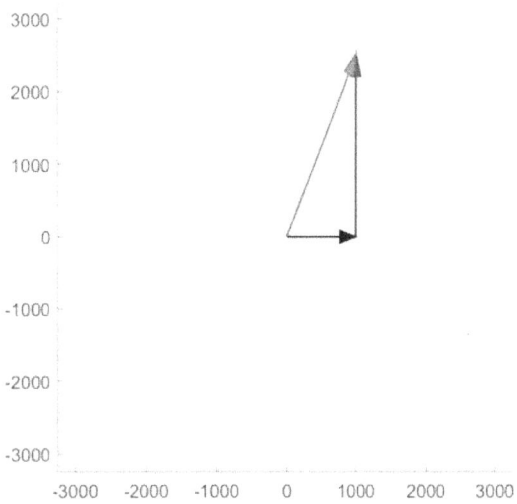

2.1.2   ADMITTANCE

The **admittance** is a measure of how easily a circuit or device will allow a current to flow. It is defined as the reciprocal of impedance, and also is presented in a complex form or a polar form:

$$Y = G + j\,B \qquad\qquad Y = |Y|\ \angle\ \theta. \qquad \text{where} \qquad \theta = tan^{-1}\left(\frac{B}{G}\right)$$

It is also the reciprocal of impedance:

$$Y = \frac{1}{Z}$$

The real part is called the conductance G, which is independent of frequency. The imaginary part is called susceptance, which is dependent on frequency. It can also be represented by a polar vector on the complex plane, where, the length of the vector gives the magnitude and the angle with reference to the horizontal axis in the anticlockwise direction. The SI unit of admittance is Siemens (symbol S) [Ref. Wikipedia].

MATLAB_Ex_2.1.2
Find the magnitude, phase angle in degrees of the admittance of the components in the example of MATLAB_Ex_2.1.1. Plot the polar and its admittance triangle on the complex plane.

```
%add the following code to the above
MATLAB code
one=coeff(1);  % an unity constant
Y=one/Z      %admittance
Y =
  phasor with properties:

    Mag: 3.6970e-04
    phase: -68.3030
clf
phplot(Y)
Current plot held
Current plot released
```

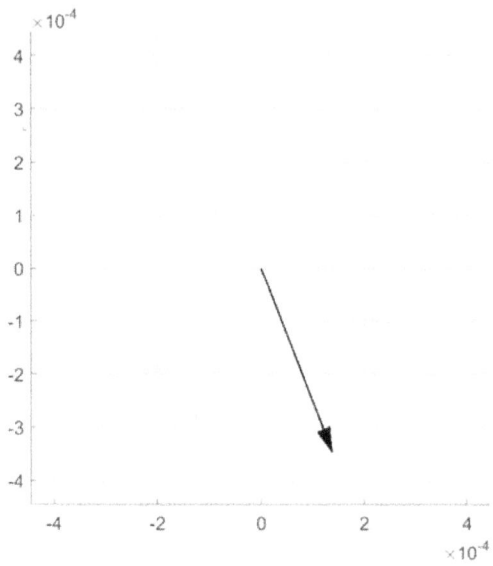

```
clf
triplot(Y)
```

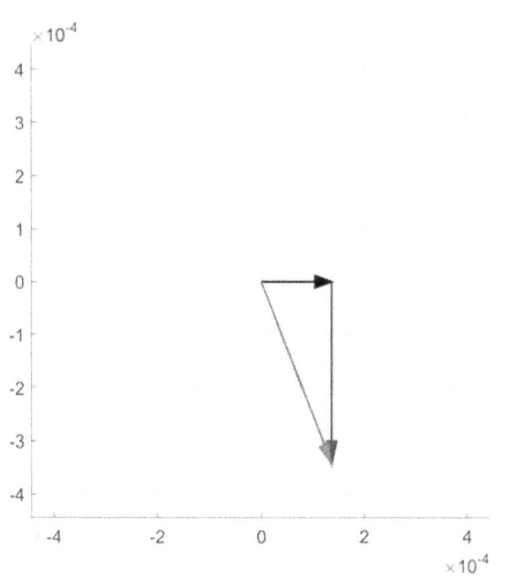

2.1.3   IMPEDANCE DIAGRAM CHANGES WITH FREQUENCY

The impedance diagram changes with frequency if it contains the inductance and /or capacitor. If the circuit contains both inductor and capacitor, the overall impedance is inductive below the resonant frequency and capacitive beyond it. At the resonant frequency, the inductive and capacitive reactances cancel and the overall impedance is purely resistive.

MATLAB_Ex_2.1.2   Find and plot the input impedance Z at a) 1000 Hz, b) 4000 Hz and

c) at the resonance frequency.

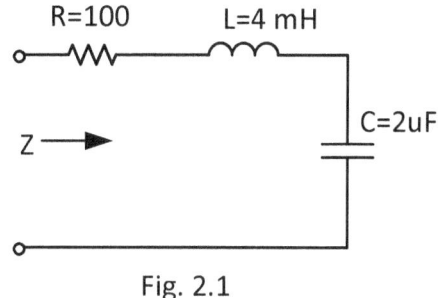

Fig. 2.1

```
%PTB2_Ex_16B.mlx
clf
R=100;  L=4e-3;  C= 2e-6;
f1=1000; f2=4000;
f0=1/(2*pi*sqrt(L*C))   %resonant frequency
f0 = 1.7794e+03
%at f1
Z1=phasor(R,(2*pi*f1*L-1/(2*pi*f1*C)), 'x2po');
triplot(Z1)
```

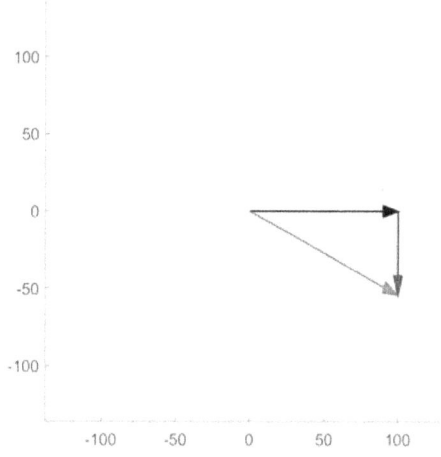

```
%at f2
clf
Z2=phasor(R,(2*pi*f2*L-1/(2*pi*f2*C)), 'x2po');
triplot(Z2)
```

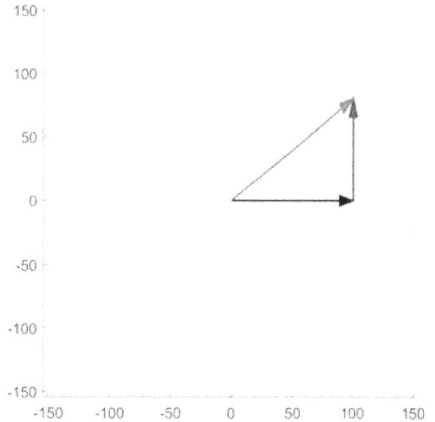

```
%at f0
clf
f0=1/(2*pi*sqrt(L*C))
f0 = 1.7794e+03
ZR=phasor(R,(2*pi*f0*L-1/(2*pi*f0*C)), 'x2po');
triplot(ZR)
```

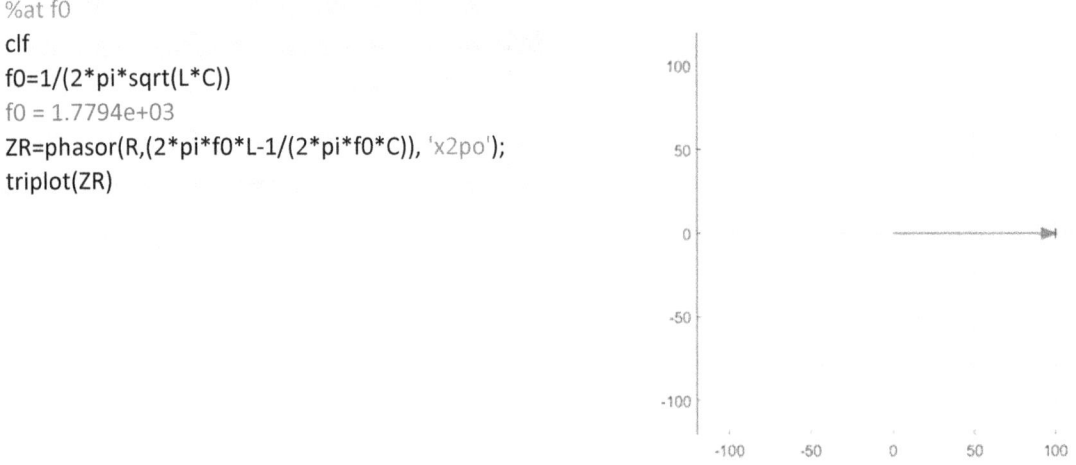

2.2   IMPEDANCE CALCULATION IN CIRCUITS

2.2.1   SERIES COMBINING OF IMPEDANCES

2.2.1.1   SERIES COMBINING OF IMPEDANCES IN COMPLEX FORM

MATLAB_Ex_2.2.1.1
Find the input impedance of 4 series connected impedances: Z1=3+j*4, Z2=2+j*3, Z3=4+j*5, and Z4=5-j*6.

```
%PTB2_Ex_5.mlx
clf
Z1=3+j*4;
Z2=2+j*3;
Z3=4+j*5;
Z4=5-j*6;
Zt=Z1+Z2+Z3+Z4
Zt =
  14.0000 + 6.0000i
ZT=phasor(real(Zt), imag(Zt), 'x2po')
ZT =
  phasor with properties:

    Mag: 15.2315
    phase: 23.1986
triplot(ZT)
```

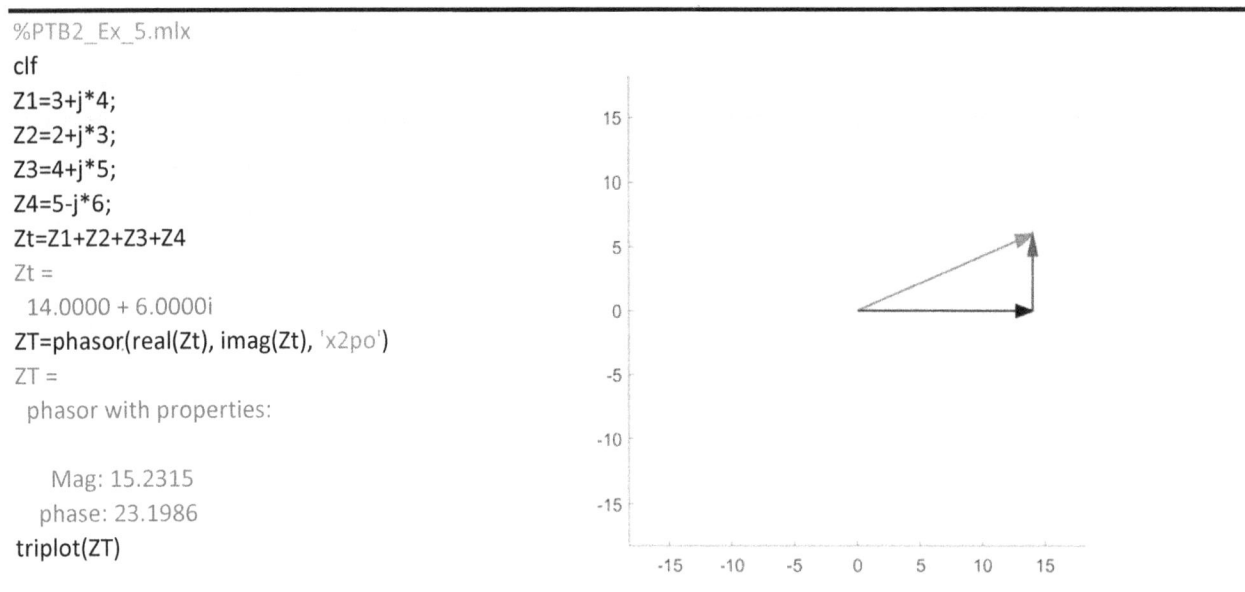

2.2.1.2 SERIES COMBINING OF IMPEDANCES IN POLAR FORM

MATLAB_Ex_2.2.1.2
Find the input impedance of 4 series connected impedances: $Z1=5 \angle30$, $Z2=10 \angle45$, $Z3=8 \angle120$ and $Z4=5 \angle-30$.
Draw the phasor diagram of all phasors and its series combination.

---

clf
%Polar form
Z1=phasor(5, 30);   Z2=phasor(10, 45); Z3=phasor(8, 120); Z4=phasor(5, -30);
ZT=Z1+Z2+Z3+Z4
ZT =
  phasor with properties:

    Mag: 18.2648
    phase: 50.0372
clf
phplot([Z1, Z2, Z3, Z4, ZT])
Current plot held
Current plot released

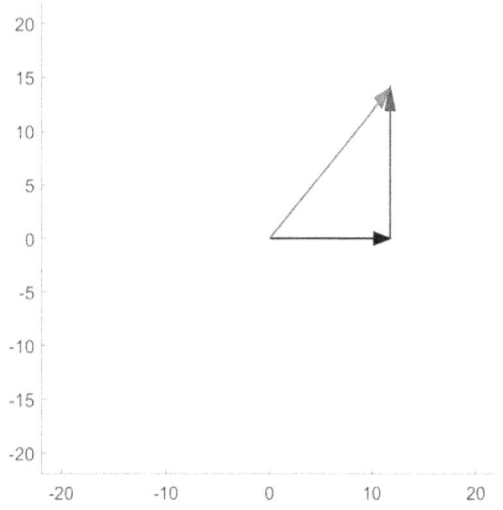

clf
triplot(ZT);

2.2.2   PARALLEL COMBINING

PHASOR TOOL BOX FUNCTION: **ZT = parallelZ**(Z, type **)**
Finds parallel combination of two or more impedances or admittances in polar or complex forms. Z is an array of either impedances (Z's) or admittances (Y's). Output of this function ZT is a polar vector for impedance or admittance.
ZT=parallelZ(Z)   to find parallel combination of all elements in an array of impedances, in polar form.
ZT=parallelZ(Z, 'cx')   to find parallel combination of all elements in an array of impedances, in complex form

2.2.2.1   PARALLEL COMBINING IN COMPLEX FORM

MATLAB_Ex_2.2.2.1
Find the input impedance of 4 parallel connected impedances: Z1=3+j*4, Z2=2+j*3, Z3=4+j*5, and Z4=5-j*6.

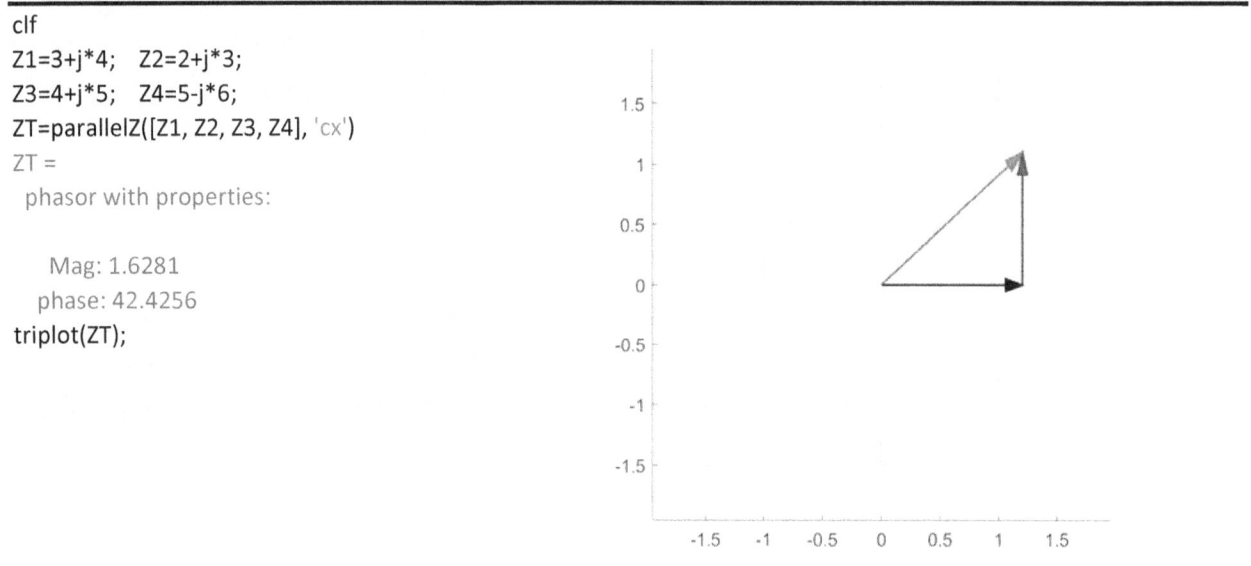

```
clf
Z1=3+j*4;   Z2=2+j*3;
Z3=4+j*5;   Z4=5-j*6;
ZT=parallelZ([Z1, Z2, Z3, Z4], 'cx')
ZT =
  phasor with properties:

    Mag: 1.6281
    phase: 42.4256
triplot(ZT);
```

2.2.2.2   PARALLEL COMBINING IN POLAR FORM

MATLAB_Ex_2.2.2.2
Find the input impedance of 4 parallel connected impedances: Z1=5 ∠30 , Z2=10 ∠45, Z3=8 ∠120 and Z4=5 ∠-30.

```
clf, clear
Z1=phasor(5, 30);   Z2=phasor(10, 45); Z3=phasor(8, 120); Z4=phasor(5, -30);
ZT=parallelZ([Z1, Z2, Z3, Z4])
ZT =
  phasor with properties:

    Mag: 2.5175
    phase: 26.7784
triplot(ZT)
```

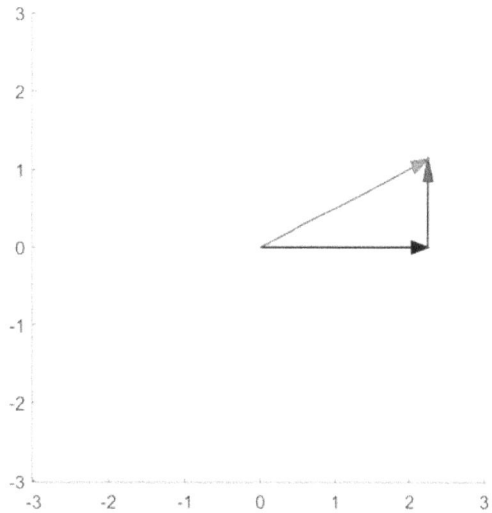

2.3   R-L-C CIRCUITS WITH INDEPENDENT VOLTAGE SOURCES

In this section we, will see the calculation and plots in complex plane and time domain of voltage, current and impedances in circuits containing R, L and C in several configurations.

2.3.1   SERIES R-L-C CIRCUITS

MATLAB_Ex_2.3.1.1
Calculate and draw the phasor diagrams, and time domain waveforms of E, I, V_R, V_L and V_C phasors in the following circuit :

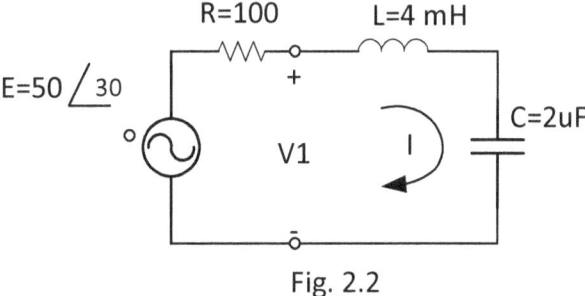

Fig. 2.2

%FTB_Ex_22.mlx
```
clf, clear
E=phasor(50, 30);
f=60;  T=1/f;
R=2; L=10e-3; C=10e-3;
```

```
XL=2*pi*f*L; XC=1/(2*pi*f*C);
Z=phasor(R, XL+XC, 'x2po');       %overall impedance in the polar form
Z1=x2ph([R, j*XL, -j*XC], 'po');  %component-wise impedances in the polar form
I=E/Z                             % current phasor
I =
  phasor with properties:

    Mag: 11.1022
    phase: -33.6350
V = [I*Z1(1), I*Z1(2), I*Z1(3)]; %voltages across all components

phplot([E, I]);
Current plot held
Current plot released
```

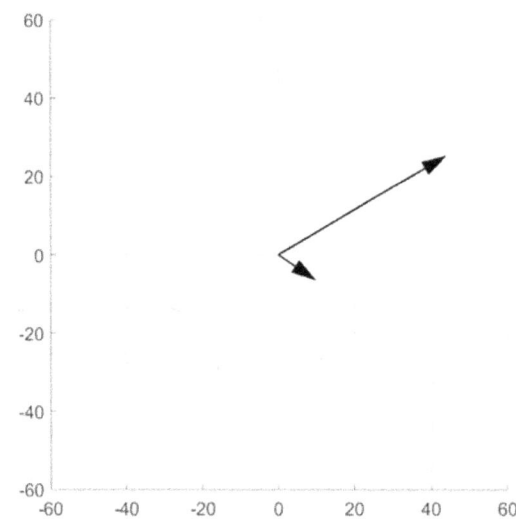

```
clf
phplot([VR, VL, VC]);
```

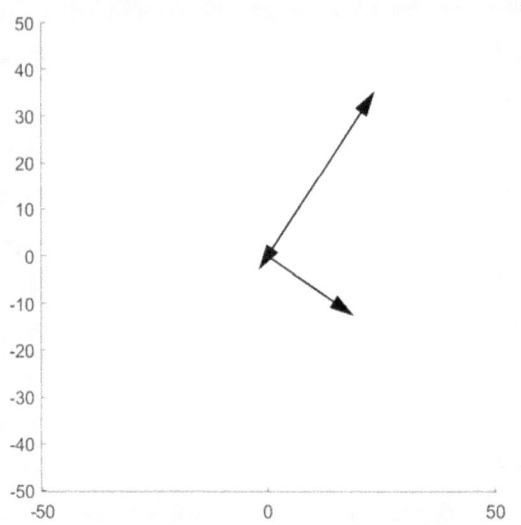

clf
phplot_signal([E, I], f, 0, 2*T);

Current plot held
Current plot released

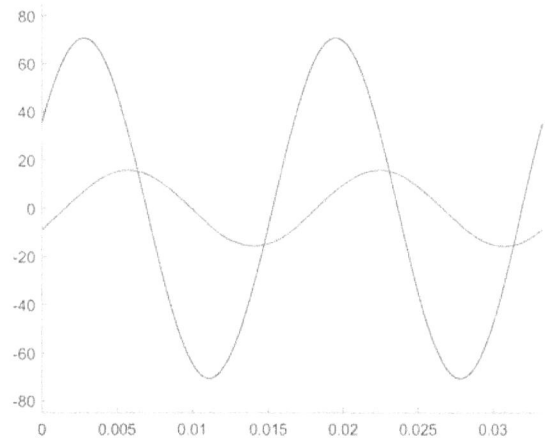

clf
phplot_signal([VR, VL, VC], f, 0, 2*T)

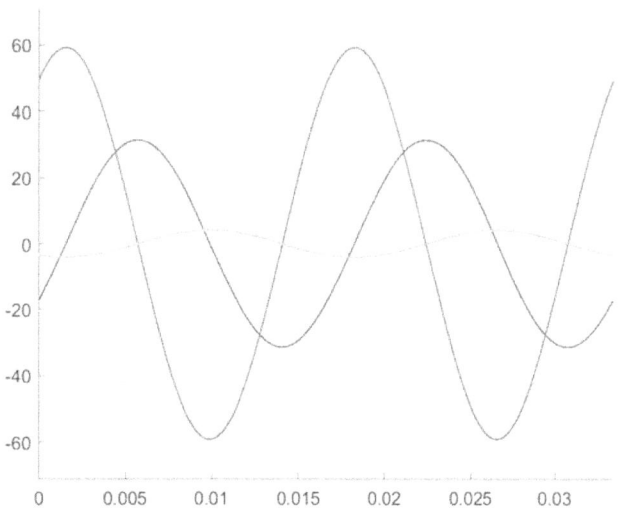

MATLAB_Ex_2.3.1.2
Calculate and plot voltage VR, VL, VC and current I
in above circuit on the complex plane.

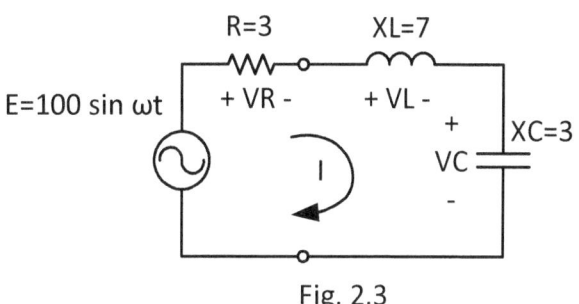

Fig. 2.3

```
%PTB2_Ex_24.mlx
clf
f=60;  T=1/f;
R=3; XL=7; XC=3;
E=phasor(100/sqrt(2), 0);
Z=R+j*(XL-XC)
Zp=x2ph(Z, 'po');
Z1=x2ph([R, j*XL, -j*XC], 'po');
V1=[E/Z1(1), E/Z1(2), E/Z1(3)];
V1(1), V1(2), V1(3)
```

ans =

  phasor with properties:

    Mag: 23.5702
   phase: 0

ans =

  phasor with properties:

    Mag: 10.1015
   phase: -90

ans =

  phasor with properties:

    Mag: 23.5702
   phase: 90

```
I=E/Zp
```

I =

  phasor with properties:

    Mag: 14.1421
   phase: -53.1301

```
phplot([E, I, V1(1), V1(2),
V1(3)]);
```

Current plot held
Current plot released

Labeling of phasors in the phasor diagram and the triangle is left for the readers to do.

2.3.2   PARALLEL R-L-C CIRCUITS

MATLAB_Ex_2.3.1.2
For the following circuit, find a) the impedance seen by the source, b) phasors I, IR, IL, IC and plot them and c) plot the signals in time domain.

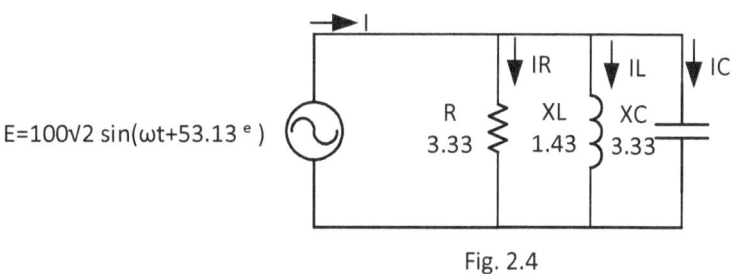

Fig. 2.4

```
%PTB2_Ex_30.mlx
clf, clear
f=60; T=1/f;
ZL=j*1.43; R=3.33; ZC=-j*3.33;
E=phasor(100 , 53.13 );
Z1=x2ph([R, ZL, ZC], 'po');        %component-wise impedances in the polar form
ZT=parallelZ(Z1)                    %overall impedance in the polar form
ZT =
  phasor with properties:
     Mag: 2.0025
     phase: 53.0337
I=E/ZT                              %overall current
I =
  phasor with properties:
     Mag: 49.9381
     phase: 0.0963
%current in each component
I1=[E/Z1(1), E/Z1(2), E/Z1(3)];
phplot([E,I,  I1(1), I1(2), I1(3)])
Current plot held
Current plot released
clf
```

phplot_signal([E, I, I1(1), I1(2), I1(3)], f, 0, T);
Current plot held
Current plot released

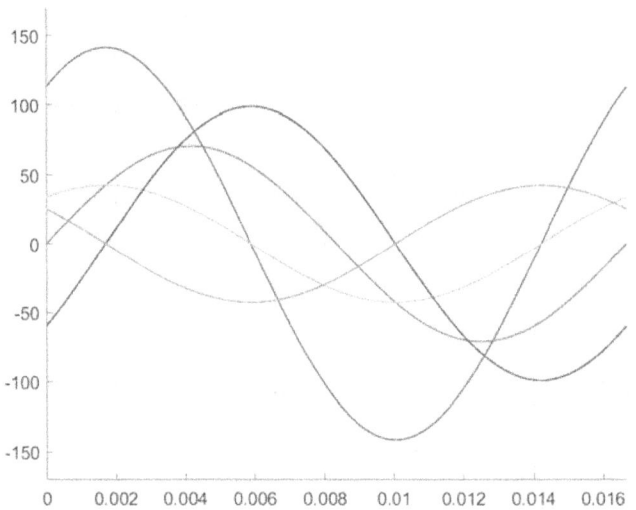

2.3.3   SERIES-PARALLEL CIRCUITS

MATLAB_Ex_2.3.1.3
Find the input impedance ZT and the phasors I, VR, VC and IC.

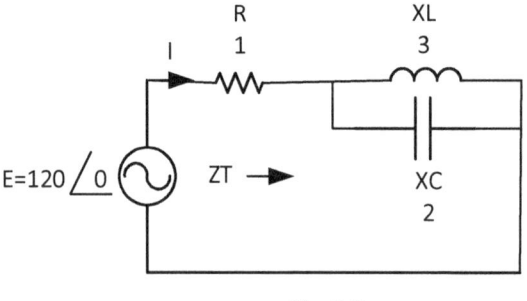

Fig. 2.5

```
R=1; XL=3; XC=2;
E=phasor(120, 0);
Z1=x2ph([R, j*XL, -j*XC], 'po');      %component-wise impedances in the polar form
Z2=parallelZ([Z1(2), Z1(3)])
Z2 =
  phasor with properties:
    Mag: 6.0000
   phase: -90
%part a)
```

```
ZT=Z1(1)+Z2;
I=E /ZT                    %overall current phasor
I =
  phasor with properties:
    Mag: 19.7279
    phase: 80.5377
%part c)
V=[I * Z1(1), I*Z2];   %V(1) is phasor voltage on R, V(2) is on the parallel combination of L and C
 %part d)
IC=V(2)/Z1(3)
IC =
  phasor with properties:
    Mag: 59.1836
    phase: 80.5377
```

MATLAB 2.3.1.4: Calculate and plot the phasor Vab.

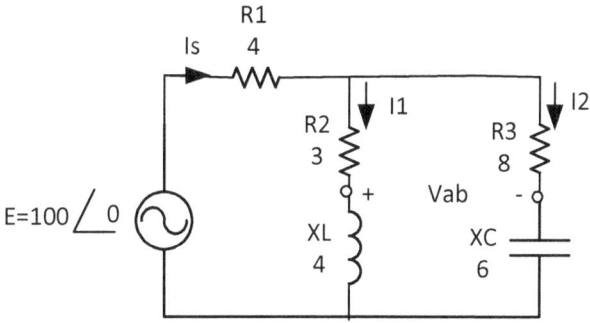

Fig. 2.6

```
%PTB2_Ex_32.mlx
R1=4; R2=3; R3=8; XL=4; XC=6;
Z1=R1; Z2=R2; Z3=j*XL; Z4=R3; Z5=-j*XC;
E=phasor(100, 0);
Z=x2ph([Z1, Z2, Z3, Z4, Z5], 'po');
ZB=parallelZ([Z(2)+Z(3), Z(4)+Z(5)])
ZB =
  phasor with properties:
    Mag: 4.4721
    phase: 26.5651
Is=E/(Z(1)+ZB)   %source current
Is =
  phasor with properties:
    Mag: 12.1268
    phase: -14.0362
V=E*(ZB/(Z(1)+ZB))
V =
```

phasor with properties:
    Mag: 54.2326
    phase: 12.5288
Va=V*Z(2)/(Z(2)+Z(3))
Va =
    phasor with properties:
    Mag: 32.5396
    phase: -40.6013
Vb=V*Z(4)/(Z(3)+Z(4))
Vb =
    phasor with properties:
    Mag: 48.5071
    phase: -14.0362
Vab=Va-Vb
Vab =
    phasor with properties:
    Mag: 24.2536
    phase: 202.8337
phplot([E, Vab])
Current plot held
Current plot released
    Mag: 12.1268
    phase: -14.0362
V=E*(ZB/(Z(1)+ZB))
V =
    phasor with properties:
    Mag: 54.2326
    phase: 12.5288
Va=V*Z(2)/(Z(2)+Z(3))
Va =
    phasor with properties:
    Mag: 32.5396
    phase: -40.6013
Vb=V*Z(4)/(Z(3)+Z(4))
Vb =
    phasor with properties:
    Mag: 48.5071
    phase: -14.0362
Vab=Va-Vb
Vab =
    phasor with properties:
    Mag: 24.2536
    phase: 202.8337
phplot([E, Vab])

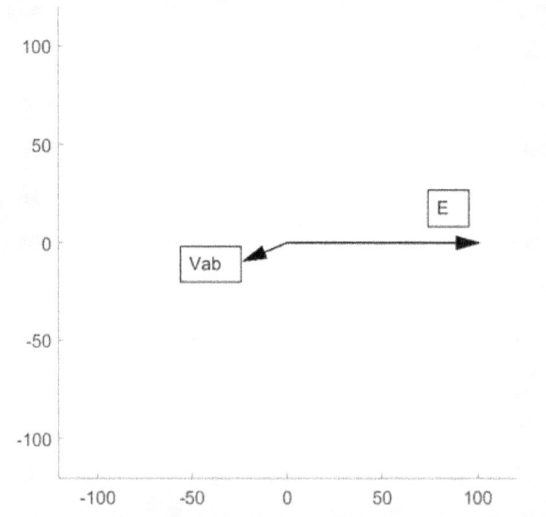

MATLAB 2.3.1.5: Calculate the input impedance ZT, and the phasors V and I.

I1=6 mA ∠20
I2=4 mA ∠0

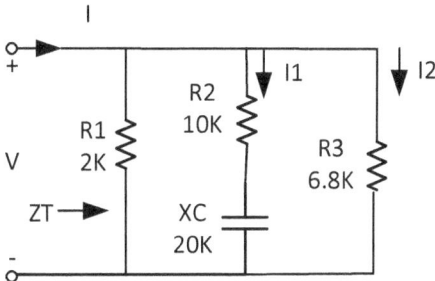

Fig. 2.8

```
%PTB2_Ex_34.mlx
clf
R1=2e3; R2=10e3; R3=6.8e3; XC=20e3;
Z1=R1; Z2=R2-j*XC; Z3=R3;
Z=x2ph([Z1, Z2, Z3], 'cx');
ZT=parallelZ([Z1, Z2, Z3], 'cx')
ZT =
  phasor with properties:
    Mag: 1.4964e+03
   phase: -3.4316
I1=phasor((6e-3), 20);
I2=phasor((4e-3), 0);
V=I1*Z(2)
V =
  phasor with properties:
    Mag: 94.8683
   phase: -43.4349
I=V/Z(1)+I1 + I2
I =
  phasor with properties:
    Mag: 0.0731
   phase: -37.0619
```

2.4   VOLTAGE DIVIDER RULE

Voltage divider rule is an important technique of circuit analysis. We will see an example of using this rule in working with phasors.

MATLAB 2.4.1:   Calculate and plot the phasor V1 in the following circuit:

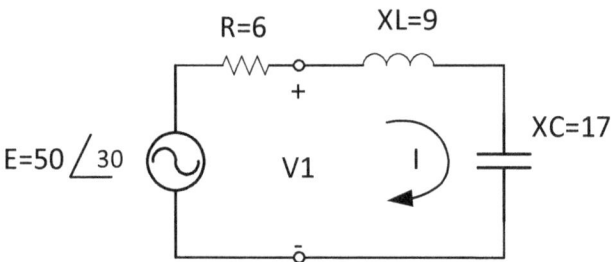

Fig. 2.14

```
%PTB2_Ex_25.mlx
%Example Fig. 15-38: R-L-C circuit
clf
ZR=phasor(6, 0); ZL=phasor(9, 90); ZC=phasor(17, -90);
ZT=ZL+ZC
ZT =
  phasor with properties:
    Mag: 8
   phase: -90
E=phasor(50, 0);
V1=E*ZT/(ZT+ZR)
V1 =
  phasor with properties:

   Mag: 40
  phase: -36.8699
phplot([E, V1])
```

2.5   CURRENT DIVIDER RULE

Like the Voltage divider rule , the current divider rule is also an important technique in analyzing ac circuits.

MATLAB 2.5.1:   Calculate the input impedance ZT, and the phasors I, I1 and I2 and plot.

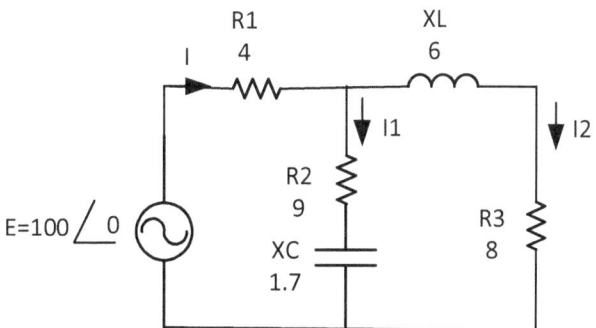

Fig. 2.15

```
%PTB2_Ex_33.mlx
R1=4; R2=9;  R3=8; XL=6; XC=7;
Z1=R1; Z2=R2-j*XC; Z3=R3+j*XL;
Z=x2ph([Z1, Z2, Z3], 'po');
E=phasor(100, 0);
%part a)
ZB=parallelZ([Z(2), Z(3)]);
Zin=Z(1) + ZB
Zin =
  phasor with properties:
    Mag: 10.6932
    phase: 1.4783
%part b)
 I=E /Zin
 I =
  phasor with properties:
    Mag: 9.3517
    phase: -1.4783
I2=I*Z(2) /(Z(2)+Z(3))      %current divider
rule
 I2 =
  phasor with properties:
    Mag: 6.2613
    phase: -35.9868
I1=I - I2
 I1 =
  phasor with properties:
    Mag: 5.4915
    phase: 38.7581
```

phplot([I, I1, I2])

2.6   WYE-DELTA CIRCUITS

AC circuits also consists of components organized as Y or Δ-connected networks. It is often required to convert one network to another for simplifying circuit configurations. The phasor tool box has two functions that take either the complex or the polar impedances /admittances.

PHASOR TOOL BOX FUNCTION: **[Z1, Z2, Z3] = delta2wye(ZA, ZB, ZC,** type **)**
Converts Δ-connected impedances to Y-connected impedances. ZA, ZB, ZC may be in the polar form or the complex form. The output Z1, Z2 and Z3 will be in the polar form only.
[Z1, Z2, Z3] = delta2wye(ZA, ZB, ZC)   input arguments are all polar impedances
[Z1, Z2, Z3] = delta2wye(ZA, ZB, ZC, 'cx')   input arguments are all complex impedances output impedances are all in polar format

This function will work similarly if the Z's are replaced by admittances Y's.

PHASOR TOOL BOX FUNCTION: **[ZA, ZB, ZC] = wye2delta(Z1, Z2, Z3,** type **)**
Converts Y-connected impedances to Δ-connected impedances. Z1, Z2, Z3 may be in the polar form or the complex form. The output ZA, ZB and ZC will be in the polar form only.
[ZA, ZB, ZC] = wye2delta(Z1, Z2, Z3)   Y-connected polar %impedances to Delta-connected polar impedances
[ZA, ZB, ZC] = wye2delta(Z1, Z2, Z3, 'po')   Y-connected complex impedances to Delta-connected polar impedances

This function will work similarly if the Z's are replaced by admittances Y's.

MATLAB 2.6.1:   Convert the Y-connected impedances as shown below to Δ-connected impedances.

Zan=-j4
Zbn=-j4
Zcn=3+j4

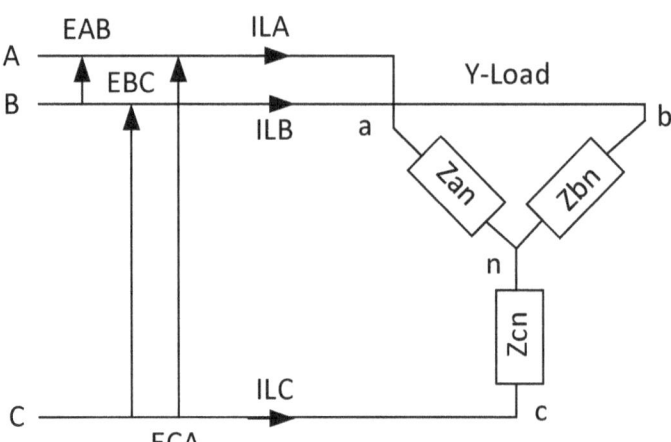

Fig. 2.16

```
%PTB2_Ex_42C.m
Zan=-j*4; Zbn=-j*4; Zcn=3+j*4;
[ZA, ZB, ZC]=wye2delta(Zan, Zbn, Zcn, 'cx')    %converting Y-impedances to Delta-impedances


ZA =
  phasor with properties:
     Mag: 7.2111
    phase: 33.6901
ZB =
  phasor with properties:
     Mag: 7.2111
    phase: 33.6901
ZC =
  phasor with properties:
     Mag: 5.7689
    phase: -109.4400
```

MATLAB 2.6.2:   Find the input impedance ZT in the following circuit:

We will illustrate one method in which the ZA, ZB and ZC (a Δ-connected network) is first converted to a Y-connected impedances Z1, Z2 and Z3. Then We find the parallel combinations of Z2+Z4 and Z3+Z5.

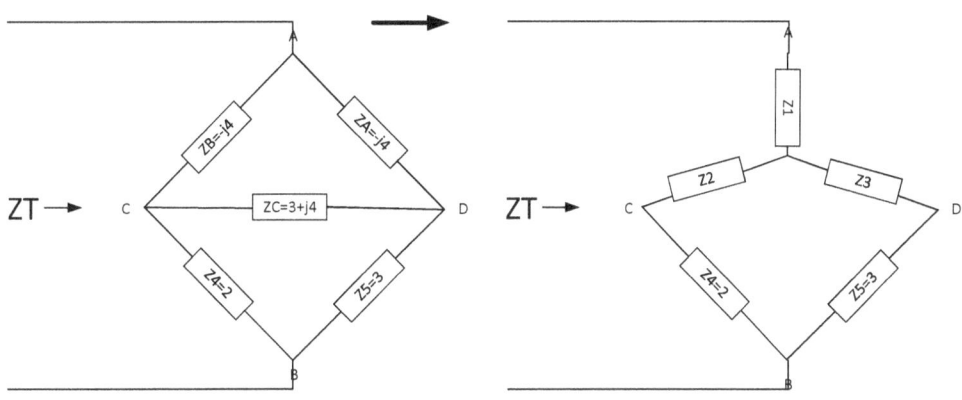

Fig. 2.17

```
%PTB2_Ex_42B.m
%all calculations are in the polar format
ZB=-j*4; ZA=-j*4; ZC=3+j*4; Z4=2; Z5=3;
Z4p=x2ph(Z4, 'cx'); Z5p=x2ph(Z5, 'cx');
[Z1, Z2, Z3]=delta2wye(ZA, ZB, ZC, 'cx');      %converting Delta impedances to Y-impedances
ZT3=parallelZ([Z1+Z4p, Z2+Z5p], 'po');
ZT=Z3+ZT3                                       %overall input impedance in the polar form
```

```
ZT =
  phasor with properties:
     Mag: 2.3455
    phase: -58.3469
```

2.7   FREQUENCY SPECTRUM OF PHASORS IN AC CIRCUITS

Following example shows how to get the frequency domain plot of voltage and current phasors in ac circuits.

MATLAB_2.7.1:   Find and plot the frequency spectrum (both the magnitude and the phase plots vs, frequency) over a frequency range of 0-40 KHz.

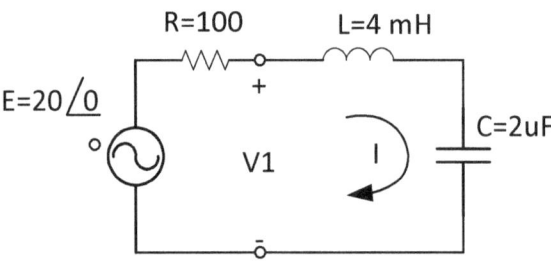

Fig. 2.18

```
%PTB2_Ex_26.mlx
clf
f=linspace(0, 40000, 400);              %range of frequencies
Ex=ph2x(phasor(20, 0));
R=100; L=4e-3; C=2e-6;
XL= 2*pi*f*L;  XC=-1./(2*pi*f*C)
Z=R+ j*(XL+XC);
V1=Ex*j*(XL+XC)./Z;                     %voltage V1 phasor
subplot(211)
plot(f, abs(V1), 'linewidth', 1 )
%magnitude of phasor V1 vs.
frequency
subplot(212)
plot(f, rad2deg(angle(V1)),
'linewidth', 1)      %phase angle
of phasor V1 vs. frequency
```

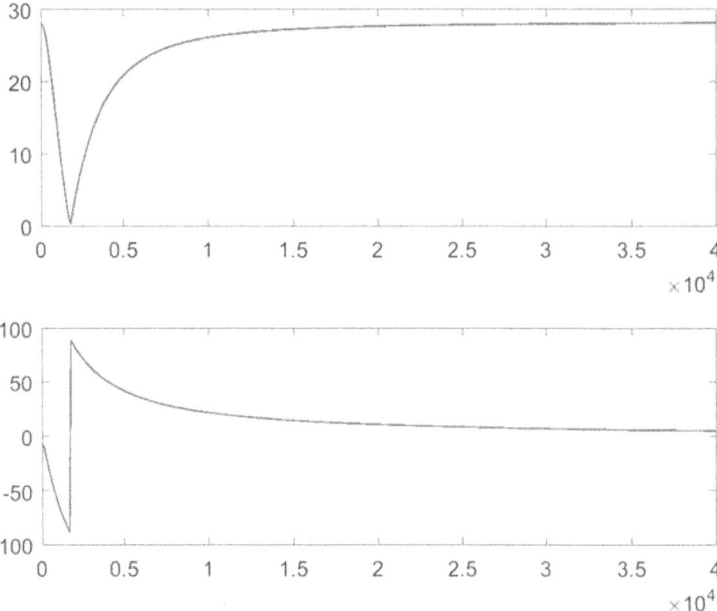

CHAPTER 3

CIRCUIT THEOREMS

LEARNING OBJECTIVES

- Mesh analysis of electrical circuits
- Nodal analysis of electrical circuits
- Determine input and output impedance
- Use superposition principles in circuit analysis
- Determine the Thevenin Equivalent circuits
- Understand and use maximum power transformer theorem.

CHAPTER-INDEX

This chapter presents the methods to analyze ac circuits such as the mesh analysis (using the Kirchoff's voltage laws), nodal analysis (using the Kirchoff's current laws). The analysis covers determining the voltage and current phasors, input and output impedances, using principles of superposition, determining the Thevenin and Norton equivalent circuits and load matching for maximum power transfer from the source to the load.

3.1   MESH ANALYSIS

3.1.1   MESH (LOOP) ANALYSIS WITH INDEPENDENT VOLTAGE SOURCES

matrix equation Z I = E
Z is the mxn matrix of Z impedances in the electrical circuit. E is
All elements are in either complex or phasor/polar form.

Case: all independent ac sources of one single frequency

Z Matrix (impedance matrix)

      Z matrix has m rows and n columns: m = # loop currents, n = m.

      All loop currents flow in the clockwise directions.

Building Z Matrix

Step 1: Enter the **series sum of all impedances in the loop** on the forward diagonal.

      Loop 1 on the element 11 (Row 1 Column 1) position

      Loop 2 on the element 22 (Row 2 Column 2) position

Step 2: Enter the **sum of impedances in the shared branch**es with negative signs in front as following:

      Branch shared between Loop j and Loop k on element jk (Row j Column k) and element kj (Row k Column j) position. Else enter zero on both element positions.

E Vector (known vector of input sources)

      E vector has m rows and 1 column: m = # loop currents.

      If the loop current enters the + terminal of a source it must be entered with a - sign in front.

      If the loop current enters the - terminal of a source it must be entered with a + sign in front.

Step 3: Enter the algebraic sum of independent voltage sources in a loop j on the j-th row of the vector.

      Loop 1 on the vector row 1 position

      Loop 2 on the vector row 2 position

I Vector (unknown vector of output loop currents)

      I vector has m rows and 1 column: m = # loop currents.

      This vector is the solution of the circuit.

      Value on the j-th row indicates value of the current in the j-th loop.

      Current in Loop 1 is on the vector row 1 position

      Current in Loop 2 is on the vector row 2 position

MATLAB_3.1.1.1:   Find the phasor currents I_1 and I_2 in the following circuit:

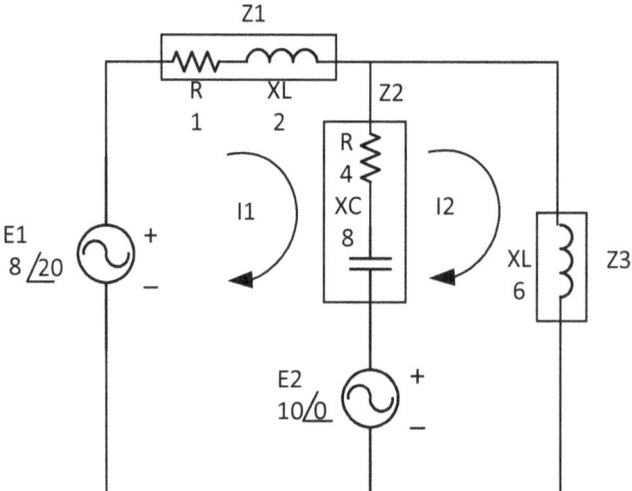

Fig. 3.1

The mesh equations are: $(Z1 + Z2)I1 \pm Z2\ I2 = E1 - E2$

$$-Z2I1 + (Z2 + Z3)\ I2 = -E2$$

The Z matrix: $Z = \begin{bmatrix} Z1 + Z2 & -Z2 \\ -Z2 & Z2 + Z3 \end{bmatrix}$

The Source Vector $E = \begin{bmatrix} E1 - E2 \\ -E2 \end{bmatrix}$

The Output Vector $I = \begin{bmatrix} I1 \\ I2 \end{bmatrix}$

```
%PTB2_Ex_40DD.mlx
%mesh analysis+input impedances seen by voltage sources
%express all E, Z and I in the phasor/polar form
Z1=phasor(1,2, 'x2po'); Z2=phasor(4,-8, 'x2po'); Z3=phasor(0,6, 'x2po');
E1=phasor(8, 20);
E2=phasor(10, 0);
Z=[Z1+Z2,   -Z2
  -Z2,    Z2+Z3 ];
E=[E1-E2; E2];      % E is a column vector
%solution for loop currents
I=Z\E;             %Solution of Z I=E equations
%solution for loop currents
I=Z\E;             %Solution of Z I=E equations
I(1)               %current I1 in phasor form
ans =
  phasor with properties:
    Mag: 1.1536
   phase: -89.1673
I(2)                %current I2 in phasor form
ans =
  phasor with properties:
    Mag: 1.0781
   phase: -53.4087
```

3.1.2   MESH ANALYSIS WITH DEPENDENT VOLTAGE SOURCES

Modifications for Dependent Voltage Source (Vx)
Step 1: Write the mesh equations from the Matrix equation **Z I = E**.

Step 2: Write the expression for Vx in terms of the independent variables.
                Example: Vx=R (I1 – I2)
Step 3: Insert the expression for Vx of step 1. Reorganize mesh equations such that all output variables I1, I2, etc.

are on the left side of the equal sign and the known sources on the right side. Re-write the modified matrix equations from the just obtained mesh equations:

Zm I = Em

Step 4: solve for vector I from the above matrix equation using

I = Zm \Em

MATLAB_3.1.2.1: Find current phasors I1 and I2.

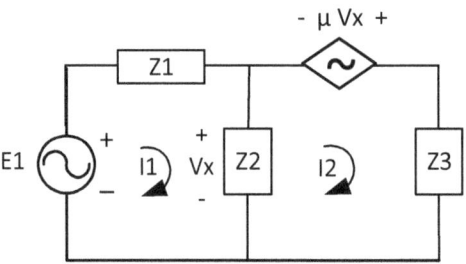

Fig. 3.2

Step 1: The Z matrix:
$$Z = \begin{bmatrix} Z1 + Z2 & -Z2 \\ -Z2 & Z2 + Z3 \end{bmatrix}$$

The Source Vector
$$E = \begin{bmatrix} E1 \\ uVx \end{bmatrix}$$

The corresponding mesh equations are:

$$(Z1 + Z2)I1 - Z2\ I2 = E1 \qquad (1)$$

$$-Z2\ I1 + (Z2 + Z3)\ I2 = uVx \qquad (2)$$

Step 2: **uVx = u Z2 [I1-I2]**

Step 3: Substituting in the mesh equation (2): $-Z2\ I1 + (Z2 + Z3)\ I2 - u\ Z2\ [I1 - I2] = 0$

$$(-Z2 - uZ2)\ I1 + (Z2 + Z3 + uZ2)\ I2 = 0$$

Step 4: The modified Z matrix is
$$Z = \begin{bmatrix} Z1 + Z2 & -Z2 \\ (-Z2 - u\ Z2) & (Z2 + Z3 + u\ Z2) \end{bmatrix}$$

and the modified Source Vector is $E = \begin{bmatrix} E1 \\ 0 \end{bmatrix}$

The Output Vector
$$I = \begin{bmatrix} I1 \\ I2 \end{bmatrix}$$

%PTB2_Ex_36A.mlx
%mesh analysis
%express all E, Z and I in the phasor/polar form
Z1=phasor(2, 0, 'x2po'); Z2=phasor(4, 0, 'x2po'); Z3=phasor(3, 0, 'x2po');
u=coeff(.5); zero=coeff(0);

```
E1=phasor(10, 0);
Z=[Z1+Z2,      -Z2
 -Z2-u*Z2,    Z2+Z3+u*Z2 ];
E=[E1 ; zero];              % E is the source vector
%solution for loop currents
I=Z\E;                     %Solution of Z I=E equations
I(1)                       %current I1 in phasor form
ans =
  phasor with properties:
    Mag: 3
   phase: 0
I(2)                       %current I2 in phasor form
ans =
  phasor with properties:
    Mag: 2
   phase: 0
```

3.1.3   MESH ANALYSIS WITH INDEPENDENT CURRENT SOURCES

Modifications for independent current Source (I=k Vx or I=k Ix)
Step 1: Open the current source and write the mesh equations.

Step 2: Write the expression for **I** in terms of the mesh currents.
                    Example: I =I3 − I2

Step 3: Insert the expression for **I** of step 2 in the mesh equations of step 1. Reorganize mesh equations such that all output variables I1, I2, etc. are on the left side of the equal sign and the known sources on the right side. Re-write the modified matrix equations from the just obtained mesh equations:
                    Zm I = Em

Step 4: solve for vector I from the above matrix equation using
                    I = Zm \Em

MATLAB_3.1.3:   Find the current phasor I1.

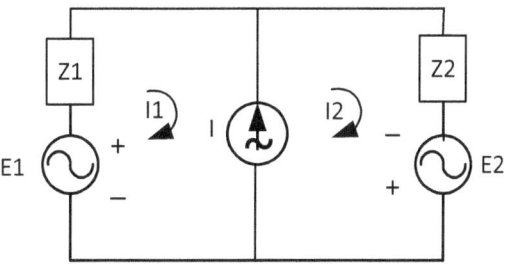

Fig. 3.3

Step 1: Open the independent current source I.

The circuit equations are:

$$Z1\ I1 + Z2\ I2\ = E1 + E2 \qquad (1)$$

Step 2:

$$-I1 + I2\ = I \qquad (2)$$

The **Zm** matrix:

$$Z = \begin{bmatrix} Z1 & Z2 \\ -1 & 1 \end{bmatrix}$$

The Source Vector **Em**

$$E = \begin{bmatrix} E1 + E2 \\ I \end{bmatrix}$$

The Output Vector

$$I = \begin{bmatrix} I1 \\ I2 \end{bmatrix}$$

```
%PTB2_Ex_37.mlx
%mesh analysis
Z1=phasor(2, 0);        Z2=phasor(4, 0);
E1=phasor(10, 0);       E2=phasor(20, 0);        I=phasor(5, 0);
one=coeff(1);
Z=[Z1,    Z2
 -one,    one ];
E=[E1+E2 ; I];          % E is the source vector
I=Z\E ;                 %Solution of Z I=E equations
I(1)                    %current I1 in phasor form
ans =
  phasor with properties:
    Mag: 1.6667
   phase: 0
```

3.1.4   MESH ANALYSIS WITH DEPENDENT CURRENT SOURCES

Modifications for dependent current Source (I=k Vx or I=k Ix):
Step 1:   Open the current source and write the mesh equations.

Step 2:   Write the expression for **I** in terms of the mesh currents.
Example: I = k1 (I1 – I2)
I = k2 (V2-V3)   →   I = k3 (I1 – I3)

Step 3:   Insert the expression for **I** of step 2 in the mesh equations of step 1. Reorganize mesh equations such that all output variables I1, I2, etc. are on the left side of the equal sign and the known sources on the right side. Re-write the modified matrix equations from the just obtained mesh equations:
Zm I = Em

Step 4: solve for vector I from the above matrix equation using
I = Zm \Em

MATLAB_3.1.4: Find the current phasor I1.

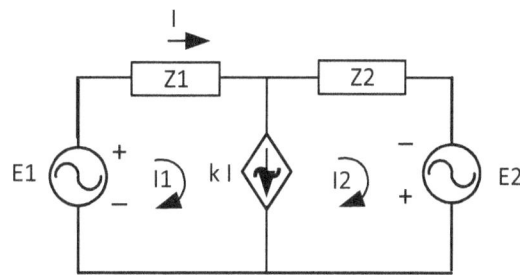

Fig. 3.4

Open the independent current source I.

The circuit equations are:

$$Z1\,I1 + Z2\,I2 = E1 + E2 \tag{1}$$

$$I1 - I2 = k\,I1 \tag{2}$$

Modified equation (2)

$$(1 - k)\,I1 - I2 = 0 \tag{2}$$

The Z matrix:

$$Z = \begin{bmatrix} Z1 & Z2 \\ 1-k & -1 \end{bmatrix}$$

The Source Vector

$$E = \begin{bmatrix} E1 + E2 \\ 0 \end{bmatrix}$$

The Output Vector

$$I = \begin{bmatrix} I1 \\ I2 \end{bmatrix}$$

```
%PTB2_Ex_38.mlx
Z1=phasor(2, 0);        Z2=phasor(4, 0);          k=coeff(.5);
E1=phasor(10, 0);       E2=phasor(20, 0);
zero=coeff(0);          one=coeff(1);
Z=[Z1,     Z2
   one-k,  -one ];
E=[E1+E2 ; zero];       % E is the source vector
%solution for loop currents
I=Z\E ;                 %Solution of Z I=E equations
I(1)                    %current I1 in phasor form
ans =
  phasor with properties:

    Mag: 7.5000
  phase: 0
```

3.2   NODAL ANALYSIS

---

matrix equation   Y V = I
All elements are in either the complex or the polar/phasor forms.

Y Matrix (admittance matrix)
> **Y matrix has m rows and n columns: m = # node voltages, n = m.**
> **All node voltages are measured with respect to the common node. The common node is generally the left bottom most node in the circuit.**
> **Each branch connected to a node must have a single admittance (for example if a branch contains two impedances Z1 and Z2, calculate Y=1/(Z1+Z2).**

Step 1:   Enter the **sum of all admittances (1/impedance, Y=1/Z) in all branches connected to a node** on the forward diagonal.
> Node 1 on the element 11 (Row 1 Column 1) position
> Node 2 on the element 22 (Row 2 Column 2) position

Step 2: enter the **sum of admittances in the shared branch**es with negative signs in front as following:
> Branch shared between Loop j and Loop k on element jk (Row j Column k) and
> element kj (Row k Column j) position. Else enter zero on both element positions.

I Vector (known vector of input sources)
> **I vector has m rows and 1 column: m = # node voltages.**
> **If a current source provides current in to the node, it must be entered with a + sign in front.**
> **If a current source draws current from the node away, it must be entered with a - sign in front.**

Step 3: enter the algebraic sum of current sources connected to a node j on the j-th row of the vector.
> Node 1 on the vector row 1 position
> Node 2 on the vector row 2 position

V Vector (unknown vector of node voltages)
> **V vector has m rows and 1 column: m = # nodes.**
> **This vector is the solution of the circuit.**
> **Value on the j-th row indicates value of the voltage on the j-th node.**
> Voltage on Node 1 is on the vector row 1 position
> Voltage on Node 2 is on the vector row 2 position

We will just take up an example of independent sources in the nodal analysis. The exercise using dependent sources is left for the readers to develop.

MATLAB_3.2.1:   Find node voltage phasors V1 and V2.

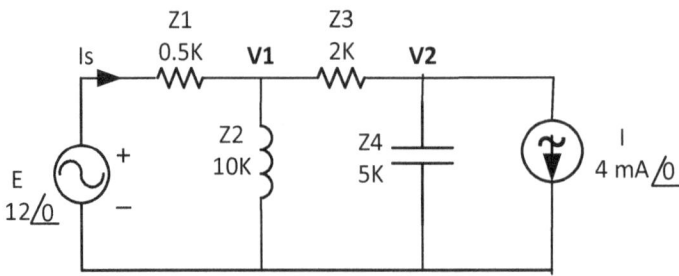

Fig. 3.5

The nodal equations are: $(Y2 + Y3)\, V1 - Y3\ V2\ \ = Is$

$$-Y3\ V1 + (Y3 + Y4)\ V2\ = -I$$

The Y matrix:

$$Y = \begin{bmatrix} Y2 + Y3 & -Y3 \\ -Y3 & Y3 + Y4 \end{bmatrix}$$

The Source Vector

$$I = \begin{bmatrix} Is \\ -I \end{bmatrix}$$

The Output Vector

$$V = \begin{bmatrix} V1 \\ V2 \end{bmatrix}$$

```
%PTB2_Ex_39.mlx
clf, clear
Z1=phasor(0.5e3, 0); Z2=phasor(10e3, 90);
Z3=phasor(2e3, 0); Z4=phasor(5e3, -90);        %all polar impedances
one=coeff(1);
Y2=one/Z2; Y3=one/Z3; Y4=one/Z4;               %converted to polar admittances
E1=phasor(12, 0);
I=phasor(4e-3, 0);
Is=E1/Z1;                                      %the source current
Y=[Y1+Y2+Y3,    -Y3
   -Y3,       Y3+Y4 ];
Ic=[Is ; -I];                                  % Ic is the Source vector
%solution for node voltages
V=Y\Ic;                                        %Solution of Y V=Ic equations
V(1),  V(2)
ans =
 phasor with properties:
    Mag: 205.9796
   phase: -42.5576
ans =
 phasor with properties:
    Mag: 185.8438
   phase: -65.9080
```

3.3   INPUT/OUTPUT IMPEDANCE IN CIRCUIT

Input impedance in a circuit is the impedance as seen between two terminals **A-A'**. It is obtained from the current that enters terminal **A** as a result of applying a voltage source across the terminals **A-A'**. This voltage source must be the only independent source in the circuit. Replace all independent sources inside the circuit by zero ("short" for voltage sources and "open" for current source).

PHASOR TOOL BOX FUNCTION: **ZTp = inputZ(**Z, V, type**)**
Finds the input impedance between two terminals **A-A'** in a circuit. Z is the mesh

impedance matrix obtained from setting all voltages sources to zero and 'opening' the current sources. Remove the component at A-A' terminals and connect an external voltage source **Vx=1.0 ∠0** across it. The positive terminal of **Vx** is connected to the terminal **A**. The **V** is a Mx1 vector containing nothing but **Vx** at the appropriate place. Output **ZTp** is a polar value. This function can also be used to find the output impedance as seen between terminals **A-A'**.

ZTp = inputZ(Z, V)   Z and V are polar matrix and phasor respectively.

ZTp = inputZ(Z, V, 'cx')   Z and V are complex matrix and complex vector respectively.

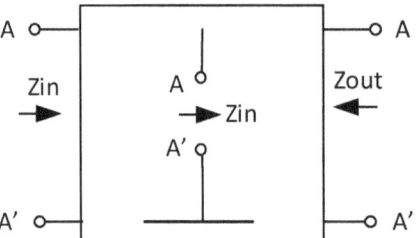

Fig. 3.6

Output impedance as seen between two terminals **A-A'** in a circuit is calculated in exactly the same way as the input impedance would have been calculated.

MATLAB Ex_3.3   Find currents I_1, I_2, and the input impedances seen by sources E_1 and E_2 in the following circuit.

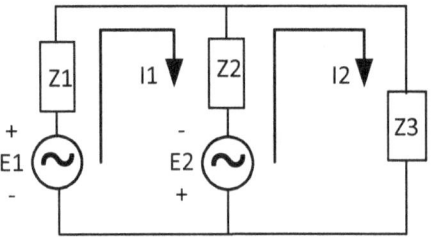

Fig. 3.7

E1=8 ∠60   E2=10 ∠0   Z1=0.5+j   Z2=4-j 8   Z3=j 6

```
clf, clear
Z1=phasor(0.5, 1, 'x2po');Z2=phasor(4, -8, 'x2po'); Z3=phasor(6, 90);
E1=phasor(8, 60);
E2=phasor(10, 0);
E=[E1+E2;  -E2];              % E is the source vector
Z=[Z1+Z2,   -Z2
 -Z2,     Z2+Z3 ];
%solution for loop currents
```

```
I=Z\E;                    %Solution of Z I=E equations
I(1)                      %current I1 in polar form
ans =
  phasor with properties:
     Mag: 1.4528
     phase: 42.0269
I(2)                      %current I2 in polar form
ans =
  phasor with properties:
     Mag: 1.1596
     phase: -39.5783
phplot([E1, I(1)]);
Current plot held
Current plot released
%input impedance as seen by source
E1
Vx=phasor(1, 0);
zero=coeff(0);
ET1=[Vx; zero];
ZT1p=inputZ(Z, ET1)
ZT1p =
  phasor with properties:
     Mag: 13.1015
     phase: 54.0048
%input impedance as seen by source
E2
ET2=[Vx; -Vx];
ZT2p=inputZ(Z, ET2)
ZT2p =
  phasor with properties:
     Mag: 8.3490
     phase: -58.4747
```

3.4   SUPERPOSITION

The principle of superposition states that the response in a circuit due to multiple sources is the sum of responses due to one source at a time.

MATLAB Ex_3.4   Find the current phasor I using the principle of superposition.

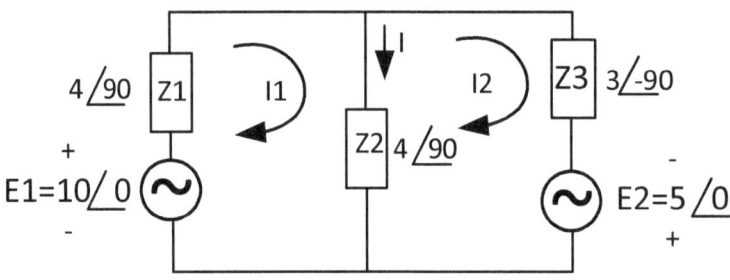

Fig. 3.8

```
%PTB2_Ex_43B.mlx
%superposition method
clf
clear
E1=phasor(10, 0);
E2=phasor(5, 0);
Z1=phasor(4, 90); Z2=phasor(4, 90); Z3=phasor(3, -90);
zero=coeff(0);  one=coeff(1);
%method 1
Z=[Z1+Z2, -Z2
   -Z2,   Z2+Z3];
% just include source E1
Es1=[E1; zero];          %source vector
I1=Z\Es1;                % solution
IL21=I1(1)-I1(2)         % current in Z2 due to one source E1 only
IL21 =
  phasor with properties:
     Mag: 3.7500
   phase: -90
% just include source E2
Es2=[zero; E2];
I2=Z\Es2 ;               % solution
IL22=I2(1)-I2(2)         % current in L2 due to one source E2 only
IL22 =
  phasor with properties:
     Mag: 2.5000
   phase: -90
IL2s=IL21+IL22           % current in L2 by superposition in the complex form
IL2s =
  phasor with properties:
     Mag: 6.2500
   phase: -90
```

3.4.1 SUPERPOSITION WITH INDEPENDENT VOLTAGE AND CURRENT SOURCES

MATLAB Ex_3.4.1   Find the current phasor I2 using the principle of superposition.

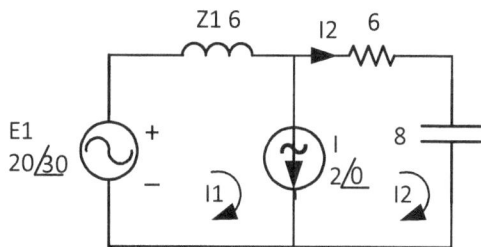

Fig. 3.9

Step 1:   Open the independent current source I.

The circuit equations are: $Z1\ I1 + Z2\ I2 = E1$ (1)

Step 2: $-I1 + I2 = I$ (2)

The modified **Zm** matrix: $Z = \begin{bmatrix} Z1 & Z2 \\ -1 & 1 \end{bmatrix}$

The Source Vector **Em** $E = \begin{bmatrix} E1 \\ I \end{bmatrix}$

The Output Vector $I = \begin{bmatrix} I1 \\ I2 \end{bmatrix}$

```
%PTB2_Ex_45B.mlx
Z1=phasor(6, 90); Z2=x2ph(6-j*8, 'po');
E1=phasor(20, 30 );              %voltage source E1 in the complex form
I=phasor(2, 0);                  %current source I in the complex form
one=coeff(1);
Z=[Z1,  Z2
   -one, one] ;                  %see the explanation in the lecture notes
%Direct method
Es=[E1; I]   ;                   %source vector
Imethod1=Z\Es;                   %solution
IR=Imethod1(2)                   %current in resistor R in the complex form
IR =
  phasor with properties:
    Mag: 4.4272
   phase: 70.2217
%Superposition method
zero=coeff(0);
Es1=[E1;zero];                   %source vector for E1 only
Is1=Z\Es1;                       %solution
```

```
Is1(2)                              %current in resistor R due source E1 only
ans =
  phasor with properties:
     Mag: 3.1623
    phase: 48.4349
Es2=[zero; I];                      %source vector for I only
Is2=Z\Es2;                          %solution
Is2(2)                              %current in resistor R sue to source I only
ans =
  phasor with properties:
     Mag: 1.8974
    phase: 108.4349
IRs=Is1(2)+Is2(2)                   %current in resistor R in the polar form obtained by the superposition method
IRs =
  phasor with properties:
     Mag: 4.4272
    phase: 70.2217
```

3.4.2   SUPERPOSITION WITH SOURCES OF DIFFERENT FREQUENCIES

MATLAB Ex_3.4.2   The sources E1 and E2 are of 100 and 200 Hz respectively. Find the current phasor I2 using the principle of superposition. Plot the time domain signals.

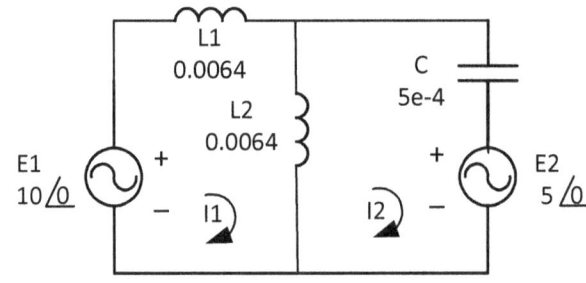

Fig. 3.10

```
%PTB2_Ex_52C.mlx
clf, clear
f1=100; f2=200;                            %two frequencies
L1=.0064; L2=L1; C=5e-4;
Z11=phasor(0, 2*pi*f1*L1, 'x2po'); Z12=Z11; Z13=phasor(0, -1/(2*pi*f1*C), 'x2po');    %impedances at f1
Z21=Z11*coeff(f2/f1);              Z22=Z21; Z23=Z13*coeff(f2/f1);                      %impedances at f2
E1=phasor(10, 0);   E2=phasor(5, 0);       %E1 and E2 in phasor form
Is1=E1/(Z11+parallelZ([Z12, Z13]));        %current from source E1 only
I2_1=Is1*Z13/(Z12+Z13)                     %by current division
I2_1 =
  phasor with properties:
     Mag: 3.3756
    phase: -90
```

```
Is2=-E2/(Z23+parallelZ([Z21, Z22])) ;        %current from source E2 only
I2_2=Is2;
[I21, t1]=phplot_signal(I2_1, f1, 0, 0.02);
Current plot held
Current plot released
```

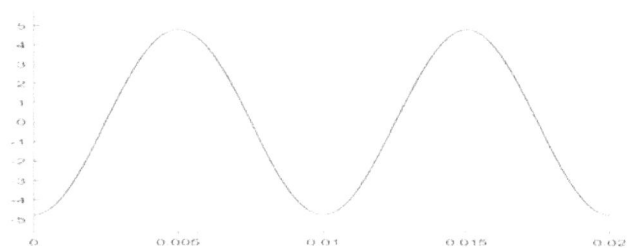

```
clf
[I22, t2]=phplot_signal(I2_2, f2, 0, .02);
Current plot held
Current plot released
```

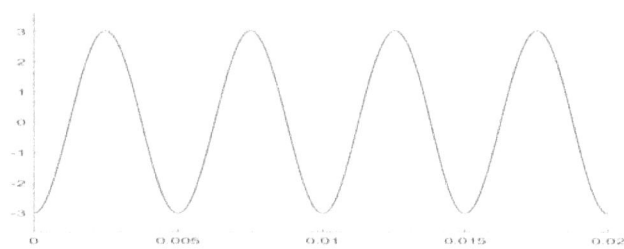

```
%superposed current
clf
plot(t1, I21+I22, 'linewidth', 1)
```

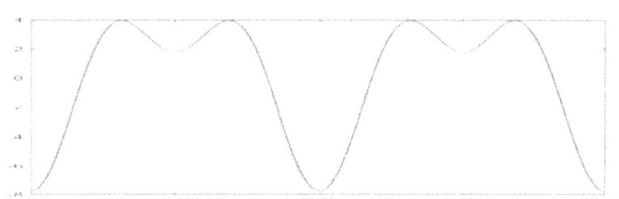

3.4.3   DC+ AC SOURCES

Circuits with dc and ac sources require a different treatment.

MATLAB Ex_3.4.3   Find and plot the waveform of the voltage V3 in the following circuit:

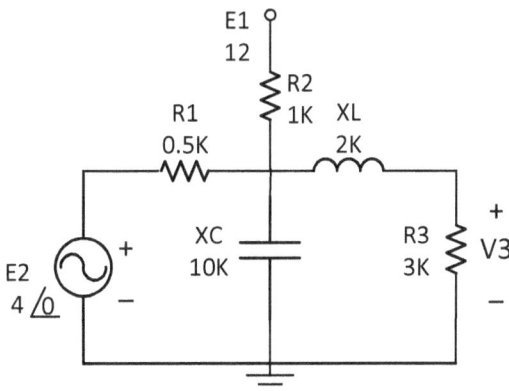

Fig. 3.11(a)

The dc equivalent circuit is shown in Fig. 3.11(b) by 'opening' capacitor, 'shorting' inductor, and 'shorting' the ac source E1. The ac equivalent circuit is shown in Fig. 3.11(c) by 'shorting' the dc source E1 to ground, which is effectively connecting R2 to the ground.

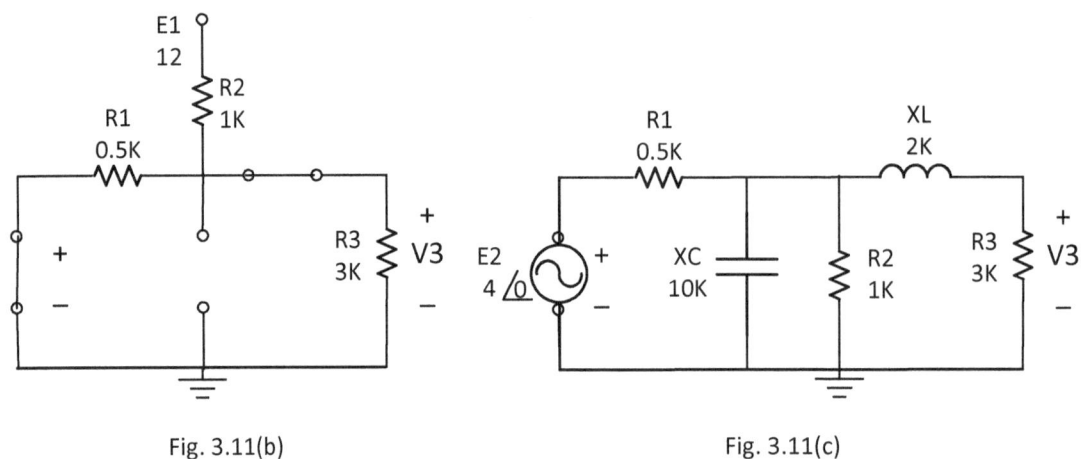

Fig. 3.11(b)   Fig. 3.11(c)

```
%PTB2_Ex_49.mlx
clf
f=60;
Z1=phasor(0.5e3, 0, 'x2po');Z2=phasor(1e3, 0, 'x2po'); Z3=phasor(0, -10e3, 'x2po');
Z4=phasor(0, 2e3, 'x2po'); Z5=phasor(3e3, 0, 'x2po');
%dc analysis
E1=phasor(12, 0);
Z15=parallelZ([Z1, Z5]);
V3dc=E1*Z15/(Z15+Z2)
V3dc =
  phasor with properties:
    Mag: 3.6000
   phase: 0
%ac signal analysis
Z23=parallelZ([Z2, Z3]);
%voltage source E2 in the polar form
E2=phasor(4, 0);
Z=[Z1+Z23, -Z23          %Z matrix
   -Z23,   Z4+Z5];
Es=[E2; coeff(0)];        %source vector
I=Z\Es;                   %solution
V3ac=Z5*I(2)
V3ac =
  phasor with properties:
    Mag: 2.5478
   phase: -43.6159
[V3act, t]=phplot_signal(V3ac, f, 0, 1/f);
```

```
V3dct=V3dc.Mag*ones(size(V3act));
clf
plot(t, V3dct+V3act, 'linewidth', 1)
grid
```

3.5   THEVENIN EQUIVALENT CIRCUIT

Thevenin equivalent circuit of Fig. 3.12(a) is shown on the right in Fig. 3.12(b). Eth is the Thevenin voltage source and Zth is the Thevenin impedance.

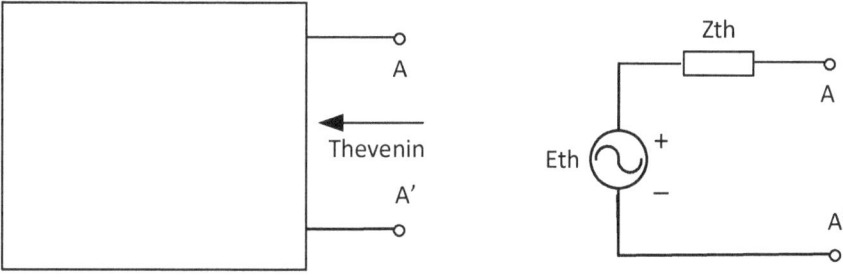

Fig. 3.12(a)   Fig. 3.12(b)

PHASOR TOOL BOX FUNCTION: **[Eth, Zth] = Thevenin(Z, E, V, varargin)**
Finds the Thevenin equivalent circuit in an electrical network between terminals **A-A'**. To achieve this, remove any component connected to terminals **A-A'**, and instead connect an external unity voltage source with zero phase **Vx. = phasor(1, 0)**. Write the impedance matrix **Z** with terminals with **A-A'** connected to **Vx** set to zero magnitude. **E** is source vector without **Vx**. **V** is the source vector containing only external voltage **Vx** across **A-A'** terminal, Positive of **Vx** connected to terminal **A**. The output **Eth** is a voltage phasor and **Zth** is the polar impedance.

[Eth, Zth] = Thevenin(Z, E, V)   Z is polar matrix, E and V are phasor vectors

[Eth, Zth] = Thevenin(Z, E, V, 'cx')   Z, E and V are all in complex form

MATLAB Ex_3.5.1   Find the Thevenin equivalent circuit as seen from terminals A-A' in the following circuit.

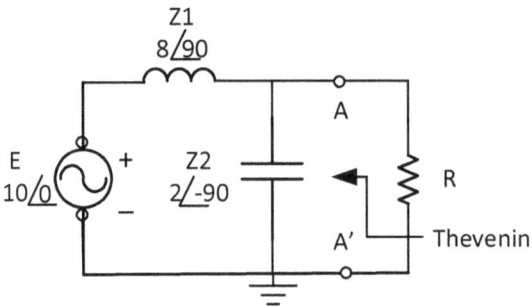

Fig. 3.12

```
%PTB2_Ex_46B.mlx
Z1=phasor(8, 90);  Z2=phasor(2, -90);
E=phasor(10, 0);                        %source vector
zero=coeff(0); Vex=phasor(1, 0);
V=[E; zero];
Z=[Z1+Z2,   -Z2
   -Z2,     Z2];                        %Z-matrix with a 'short' between terminals A and A'
Vx=[zero; -Vex];
[ Vth,  Zth] = Thevenin( Z, V, Vx, 'po' )
Vth =
  phasor with properties:
     Mag: 3.3333
   phase: 180
Zth =
  phasor with properties:
     Mag: 2.6667
   phase: 270
```

MATLAB Ex_3.5.2   Find the Thevenin equivalent circuit as seen from terminals A-A'.

Fig. 3.13

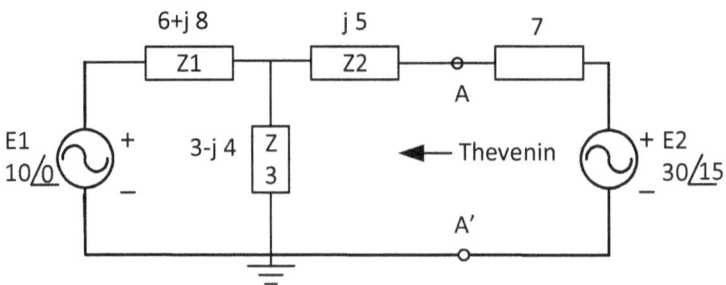

```
%PTB2_Ex_48B.mlx
%Ex19.8
clf
```

```
Z1=x2ph(6+j*8, 'po');  Z2=x2ph(3-j*4, 'po') ; Z3=phasor(5, 90);
E1=phasor(10, 0);
E2=phasor(30, 15);
zero=coeff(0);
E=[E1;zero];                %source vector
Vex=phasor(1, 0);
Vx=[zero; -Vex];
Z=[Z1+Z2,     Z2
  -Z2,       Z2+Z3];       %Z-matrix with a 'short' between terminals a and a'
[ Vth,  Zth] = Thevenin( Z, E, Vx, 'po' )
Vth =
  phasor with properties:
    Mag: 5.0767
   phase: 282.9074
Zth =
  phasor with properties:
    Mag: 1.6529
   phase: 325.4179
```

MATLAB Ex_3.5.3   Find the Thevenin equivalent circuit as seen from terminals A-A'.

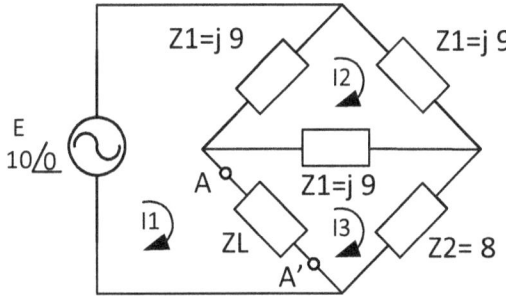

Fig. 3.14

```
%PTB2_Ex_51C.m
clf
clear
Z1=phasor(9, 90); Z2=phasor(8, 0);
E=phasor(10, 0);
zero=coeff(0);
%Write Z matrix with the terminals A-A' are shorted
V=[E; zero; zero];
Vex=phasor(1, 0);
Vx=[-Vex; zero; Vex];
Z= [ Z1,       -Z1,            zero
    -Z1,       Z1+Z1+Z1,       -Z1
    zero,      -Z1,            Z1+Z2];
```

[Eth, Zth]=Thevenin(Z, V, Vx, 'po')

Eth =
  phasor with properties:
    Mag: 8.5440
    phase: 343.6861
Zth =
  phasor with properties:
    Mag: 5.5073
    phase: 82.4879

3.6   MAXIMUM POWER TRANSFER THEORM

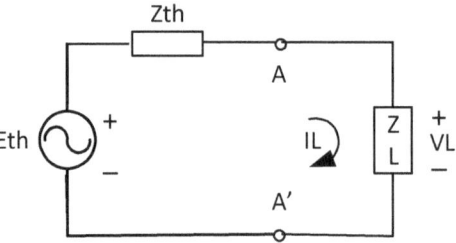

Fig. 3.15

Apparent power delivered to a load impedance ZL is given by

$$SL = VL * conj(IL) = Eth\frac{ZL}{Zth + ZL}\ \ conj\left(\frac{Eth}{Zth + ZL}\right)$$

where VL and IL are phasors and Zth and ZL are polar impedances.

The maximum power is delivered to the load when

$$ZL = conjugate(Zth)$$

When this condition exists, we say that the load is matched to the source.

The maximum Power delivered to the load is calculated from,

$$PLmax = SL.Mag * cosd(SL.phase)$$

We will use a **PHASOR TOOL BOX FUNCTION: [S, P, Q, Fp, ph] = PWR**(V, I, type) to calculate the maximum power delivered to the load. This function is explained in the Chapter 4 in detail. **P** is the maximum power delivered to the load, Fp is the power factor and **ph** indicates whether the power factor is lagging or leading type.

MATLAB Ex_3.6.1   Find the Load impedance ZL for maximum power transfer and the amount of maximum real power delivered and the power factor to the load..

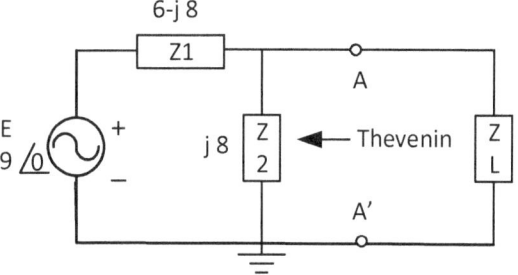

Fig. 3.16

```
%PTB2_Ex_53B.mlx
clf
clear
Z1=x2ph(6-j*8, 'po');  Z2=x2ph(j*8, 'po');
E=phasor(9, 0);
zero=coeff(0);
Z=[Z1+Z2,      -Z2
   -Z2,        Z2 ];
V=[E; zero];                      %Source vector
Vex=phasor(1, 0);                 %external voltage source for calculating Thevenin circuit
Vx=[zero; -Vex];                  %Source vector with E=0 and the external voltage Vex at the A-A' terminals
[Eth, Zth]=Thevenin(Z, V, Vx, 'po')
Eth =
  phasor with properties:
    Mag: 12
    phase: 90
Zth =
  phasor with properties:
    Mag: 13.3333
    phase: 36.8699
ZL=conj(Zth)                      %The load impedance for maximum power transfer
ZL =
  phasor with properties:
    Mag: 13.3333
    phase: -36.8699
VL=Eth*ZL/(Zth+ZL)                %voltage across the load
VL =
  phasor with properties:
    Mag: 7.5000
    phase: 53.1301
IL=Eth/(Zth+ZL)                   %current in the load
IL =
  phasor with properties:
    Mag: 0.5625
```

phase: 90
%max power delivered to the load
[Ss, Pp, Qq, Fp, ph]=PWR(VL, IL, 'ph')   %maximum power in the load
Ss =
  phasor with properties:
     Mag: 4.2187
     phase: -36.8699
Pp =
  phasor with properties:
     Mag: 3.3750
     phase: 0
Qq =
  phasor with properties:
     Mag: 2.5312
     phase: -90
Fp = 0.8000
ph = 'leading'

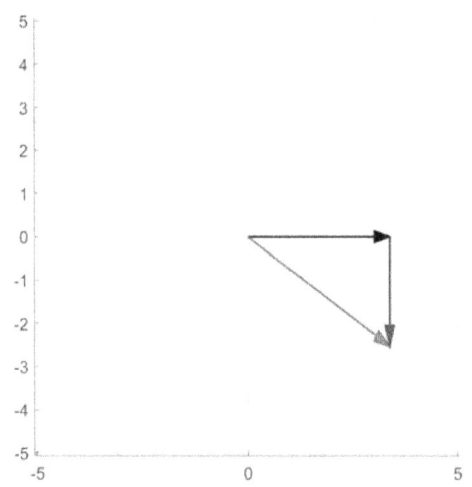

MATLAB Ex_3.6.2   Find the Load impedance ZL for maximum power transfer and the amount of the maximum
                         real power delivered and the power factor to the load.

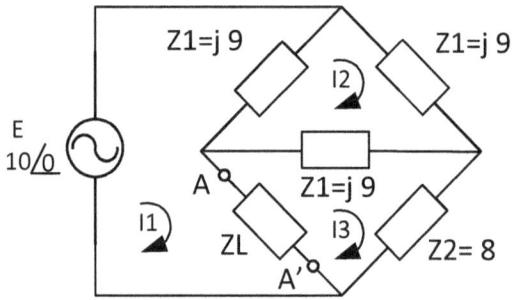

Fig. 3.17

```
%PTB2_Ex_54B.mlx
clf
clear
Z1=phasor(9, 90); Z2=phasor(8, 0);
E=phasor(10, 0);
zero=coeff(0); one=coeff(1);
```
%Write Z matrix with the terminals A-A' are shorted
```
V=[E; zero; zero];                    % source vector
Vex=phasor(1, 0);                     %external voltage source for calculating Thevenin circuit
Vx=[-Vex; zero; Vex];                 %Source vector with E=0 and the external voltage Vex at the A-A' terminals
Z= [ Z1,       -Z1,         zero
    -Z1,    Z1+Z1+Z1,       -Z1
```

```
    zero,       -Z1,                 Z1+Z2];
[Eth, Zth]=Thevenin(Z, V, Vx)
Eth =
  phasor with properties:
     Mag: 8.5440
   phase: 343.6861
Zth =
  phasor with properties:
     Mag: 5.5073
   phase: 82.4879
```

ZL=conj(Zth)   %The load impedance for maximum power transfer

```
ZL =
  phasor with properties:
     Mag: 5.5073
   phase: -82.4879
```

VL=Eth*ZL/(Zth+ZL)   %voltage across the load

```
VL =
  phasor with properties:
     Mag: 32.6765
   phase: 261.1983
```

IL=Eth/(Zth+ZL)   %current in the load

```
IL =
  phasor with properties:
     Mag: 5.9333
   phase: 343.6861
```

%max power delivered to the load

[Ss, Pp, Qq, Fp, ph]=PWR(VL, IL)   %maximum power in the load

```
Ss =
  phasor with properties:
     Mag: 193.8805
    phase: -82.4879
Pp =
  phasor with properties:
     Mag: 25.3472
    phase: 0
Qq =
  phasor with properties:
     Mag: 192.2164
    phase: -90
Fp = 0.1307
ph = 'leading'
```

CHAPTER 4

AC POWER

LEARNING OBJECTIVES

- Apparent power, real power and reactive powers in R-L_C circuits.
- Find the power factor.
- Draw the triangular plot of powers.

CHAPTER INDEX

4.1   DEFINITIONS

Consider an electrical circuit with an ac source V connect at the input as shown below:

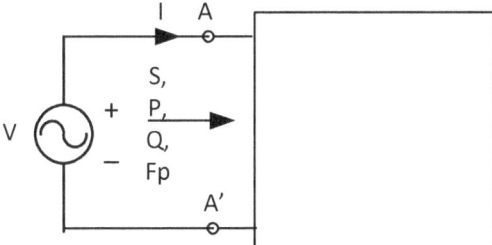

Fig. 4.1a

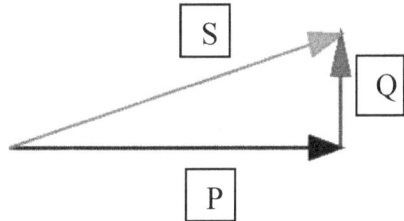

Fig. 4.1b

The apparent, the real, reactive powers delivered by the source to the circuit are defined as below:

Apparent Power   S = V I*   Volt Amp (VA)

      MATLAB: $S = \frac{1}{2}V * conj(I)$ V, I : Voltage, current as complex vectors

      Phasor Tool Box: $S = V * conj(I)$ V, I : Voltage, current as phasors

Real Power   P = V I cos(θ)   Watt (W)

      MATLAB: $P = real(\frac{1}{2}V * conj(I))$ V, I : Voltage, current as complex vectors

      Phasor Tool Box: $P = real(V * conj(I))$ V, I : Voltage, current as phasors

Reactive Power   Q = V I sin(θ)   VAR

      MATLAB: $Q = imag(\frac{1}{2}V * conj(I))$ V, I : Voltage, current as complex vectors

      Phasor Tool Box: $P = imag(V * conj(I))$ V, I : Voltage, current as phasors

Power Factor $Fp = \frac{P}{|S|} = \cos(\theta)$ Leading or Lagging

      θ is considered positive if measured in counter clockwise direction. Positive angle gives **Lagging** power factor. Negative angle gives **Leading** power factor.

MATLAB: $Fp = P/abs(S)$ V, I : Voltage, current as complex vectors
Phasor Tool Box: $Fp = P/(S.Mag)$ V, I : Voltage, current as phasors

The generator generates S power in order for the customers to consume P power. The generators bills for the S power. Therefore, the ideal value of the power factor P/|S| is unity. Normally the power factor is less than unity and special efforts are needed to bring it up to unity.

PHASOR TOOL BOX FUNCTION: **[S, P, Q, Fp, ph] = PWR(V, I, type)**
Calculates the power crossing two terminals A-A' in an electrical circuit. **V** and **I** are respectively the rms voltage and rms current phasors in the circuit at the terminals **A-A'**. **S** is apparent power, **Q** is reactive power and **P** is real power (not in phasor form), Fp is the power factor and **ph** indicates whether the power factor is lagging or leading type. The **type** indicates the type of input arguments. All output variables are in polar format. It also displays a triangle plot of S, P and Q polars.

[S, P, Q, Fp, ph] = PWR(V, I)   V and I are phasors
[S, P, Q, Fp, ph] = PWR(V, I, 'cx')   V and I are complex vectors

4.2   EXAMPLES OF POWER CALCULATIONS IN SINGLE PHASE AC CIRCUITS

MATLAB Ex_4.2.1   Find the total real, reactive and apparent power in the following network.

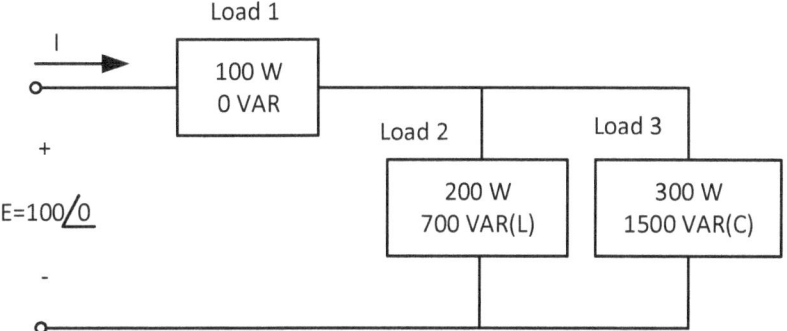

Fig. 4.2

```
%PTB2_Ex_90.m
%examples 20.3
clf, clear
%Load 1
P1=200; Q1=j*0;
%Load 2
P2=500; Q2=j*1200;
%Load 3
P3=200; Q3=-j*900;
%Total power
PT=P1+P2+P3    %total wattage in the complete circuit
PT = 900
QT=Q1+Q2+Q3   %total VAR in the
complete circuit
QT =
   0.0000e+00 + 3.0000e+02i
ST=PT+QT    %total apparent power  VA
ST =
   9.0000e+02 + 3.0000e+02i
STm=abs(ST)  %magnitude of ST
STm = 948.6833
Fp=PT/STm    %power factor
Fp = 0.9487
STp=x2ph(ST, 'po');
triplot(STp)
```

MATLAB Ex_4.2.2   Find the voltage VR, VL, real, reactive and apparent power being delivered by the source E in the following network. Use frequency of 1 Hz.

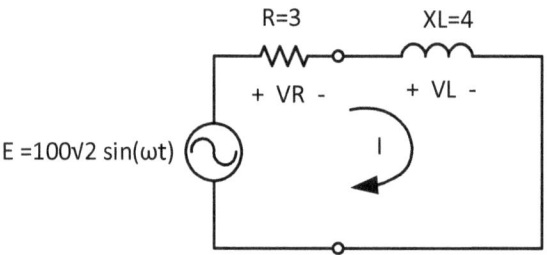

Fig. 4.3

```
%PTB2_Ex_23.mlx
clf
clear
f=1;  T=1/f;
ZR=phasor(3, 0);  ZL=phasor(4, 90);
ZT=ZR+ZL;
E=phasor(100, 0);
I=E/ZT;
VR=I*ZR
VR =
  phasor with properties:
    Mag: 60
   phase: -53.1301
VL=I*ZL
VL =
  phasor with properties:
    Mag: 80
   phase: 36.8699
phplot([VR, VL, E]);
Current plot held
Current plot released
Clf
phplot_signal([VR, VL, E], f, 0, T);
Current plot held
Current plot released
clf
[S, P]=PWR(E, I)
S =
  phasor with properties:
    Mag: 2000
   phase: 53.1301
P =
  phasor with properties:
```

Mag: 1200
phase: 0

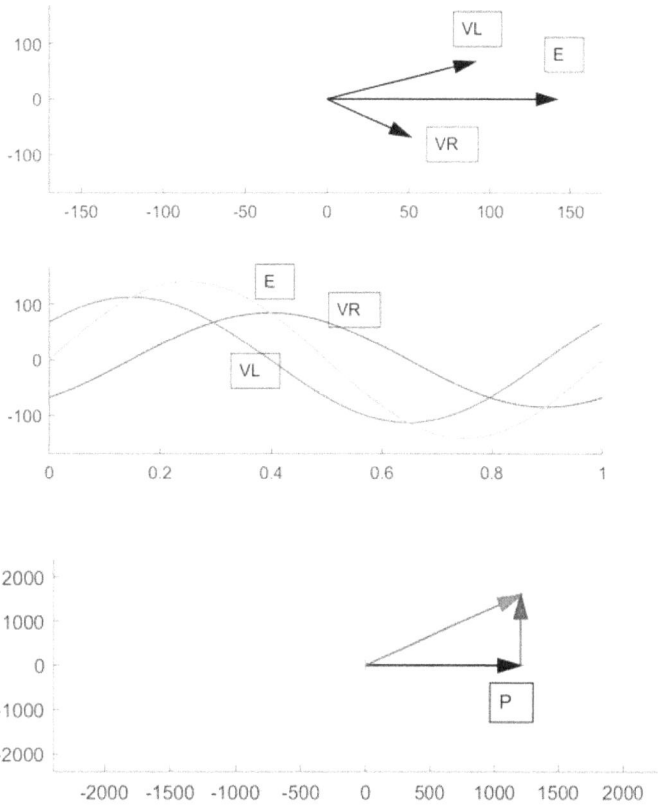

MATLAB Ex_4.2.3   Find the total real, reactive and apparent power in the following network. Use frequency of 1 Hz.

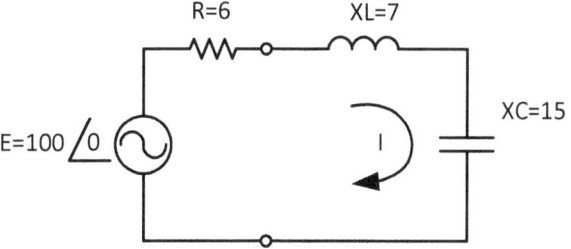

Fig. 4.4

%PTB2_Ex_55B.m
clf
clear
E=phasor(100, 0);

```
ZR=phasor(6, 0); ZL=phasor(7, 90); ZC=phasor(15, -90);
ZT=ZR+ZL+ZC;
I=E/ZT;
S=E*conj(I)
```
S =
  phasor with properties:
    Mag: 1.0000e+03
    phase: -53.1301
%Direct calculation and power triangle
[S, P, Q, Fp, ph]=PWR(E, I)
S =
  phasor with properties:
    Mag: 1.0000e+03
    phase: -53.1301
P =
  phasor with properties:
    Mag: 600.0000
    phase: 0
Q =
  phasor with properties:
    Mag: 800.0000
    phase: -90
Fp = 0.6000
ph = 'leading'

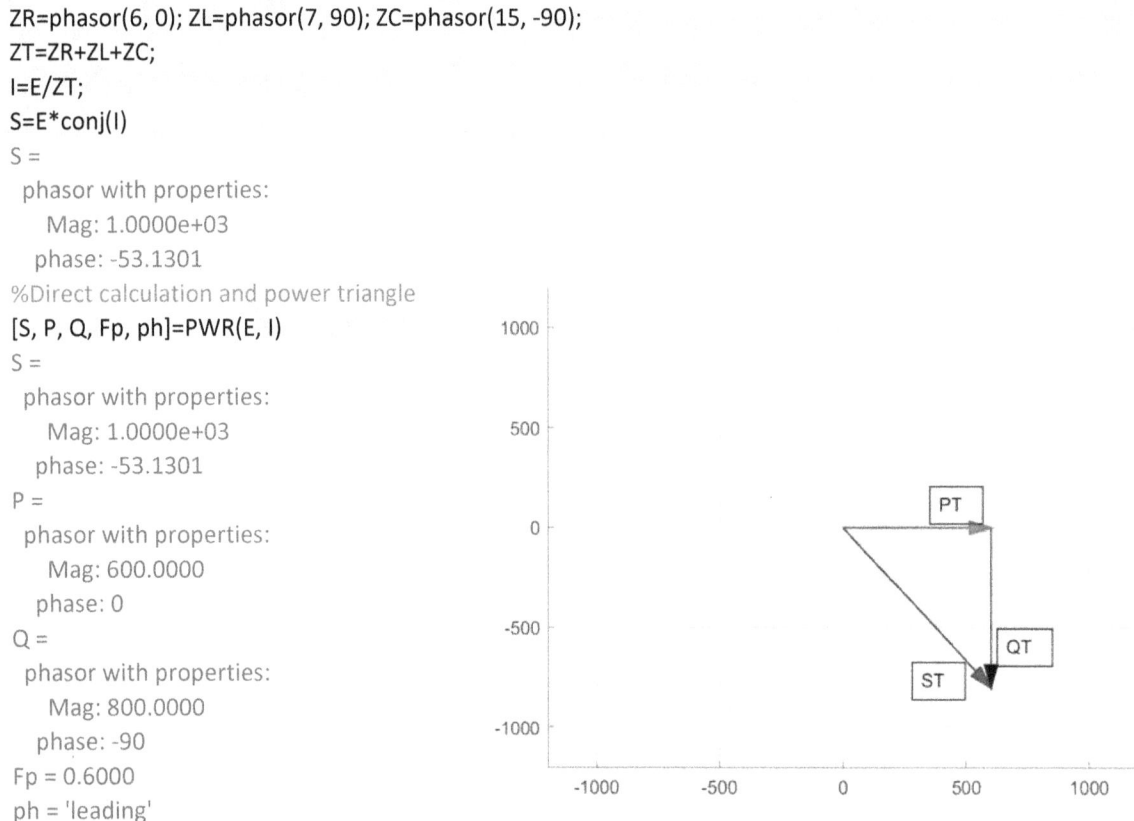

CHAPTER 5

TRANSFORMER

LEARNING OBJECTIVES

- Transformer equations and the equivalent circuit
- Calculation of impedances, voltage and current phasors

CHAPTER INDEX

5.1 TRANSFORMER EQUATIONS

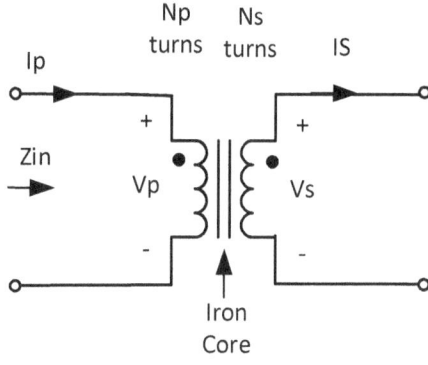

Fig. 5.1a

Turn Ratio	$a = \frac{Np}{Ns}$

Turn Ratio $\qquad a = \frac{Np}{Ns}$

Secondary side voltage $\qquad Vs = a\, Vp$

Power transfer is lossless $\qquad Vp\, Ip = Vs\, Is$

Secondary side current $\qquad Is = \left(\frac{1}{a}\right) Ip$

Impedance transfer from the
secondary winding to the primary winding $\quad Zp = a^2 Zs$

The equivalent circuit at medium frequencies of interest in electrical circuits (with 95% efficiency and accuracy) is given in Fig. 5.1b)

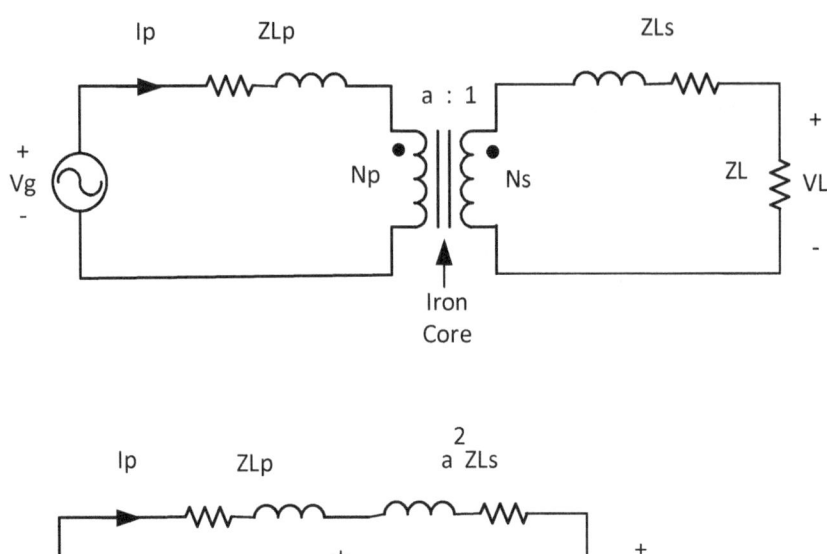

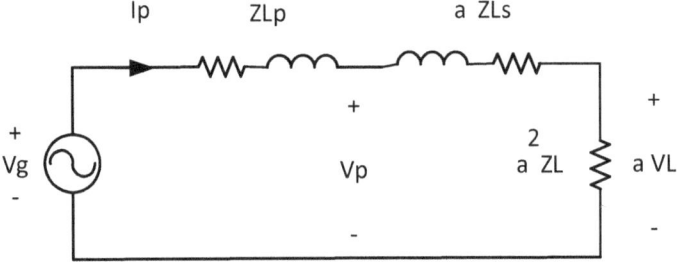

Fig. 5.1b

ZLp is the sum of the winding resistance and the leakage inductance of the primary winding at the frequency of operation. ZLs is the sum of the winding resistance and the leakage inductance of the secondary winding at the frequency of operation. The impedance ZL on the secondary winding transfers on the primary side as a^2 ZL.
The voltage VL on the secondary side transfers as a VL. The resulting equivalent circuit is analyzed by conventional or phasor methods.

5.2 TRANSFORMER EXAMPLES

MATLAB Ex_5.1   In the transformer circuit shown below, the current in the secondary winding is measured to be 100 Ma $\angle$0. Find the a) primary current, b) the voltage across the load , VL, c) the source voltage Vg and d) input impedance Zin as seen on the primary winding.

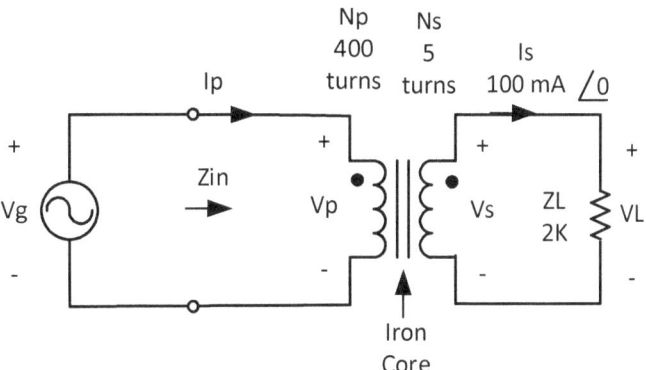

Fig. 5.2

```
%PTB_Ex_81.m
%Ch23_3
Np=40; Ns=5;                        %primary and secondary turns
a=coeff(Np/Ns);                     %Turns ratio Np/Ns
ainv=coeff(Ns/Np);                  %reverse turn ratio Ns/Np
a2=coeff((Np/Ns)^2);                %square of turns ratio
Is=phasor(100e-3, 0); ZL=phasor(2e3, 0);   %given info
%part a)
Ip=ainv*Is
Ip =
   phasor with properties:
     Mag: 0.0125
     phase: 0
VL=Is*ZL                            %voltage on the secondary winding
VL =
   phasor with properties:
     Mag: 200
     phase: 0
Vg=a*VL                             %voltage on the primary winding
Vg =
   phasor with properties:
     Mag: 1600
     phase: 0
Zin=a2*ZL                           %impedance ZL as reflected on the secondary side
Zin =
   phasor with properties:
     Mag: 128000
     phase: 0
```

MATLAB Ex_5.2   The circuit below shows the residential utility wiring in United States. Every residence is connected to only one phase of the 3-phase Y-connected utility. Thus, every residence effectively gets 240 V supply, but divided in two 120 Vac supplies A and B. with respect to the center-tap line. The residential load is further distributed over these two 120 Vac sources A and B in a 'balanced' manner. The figure shows that ten bulbs of 60 W each and the 200 W TV are connected to the supply A while the 2000 W Air Conditioner is connected to full 240 V supply. Find a) the current in the primary winding I_P, and the effective impedance Z_{in} as seen into the primary winding, b) the currents I_1 and I_2 , c) the line-line voltage on the utility side and d) the total 3-phase power drawn from the utility assuming that all 3-phases are load balanced.

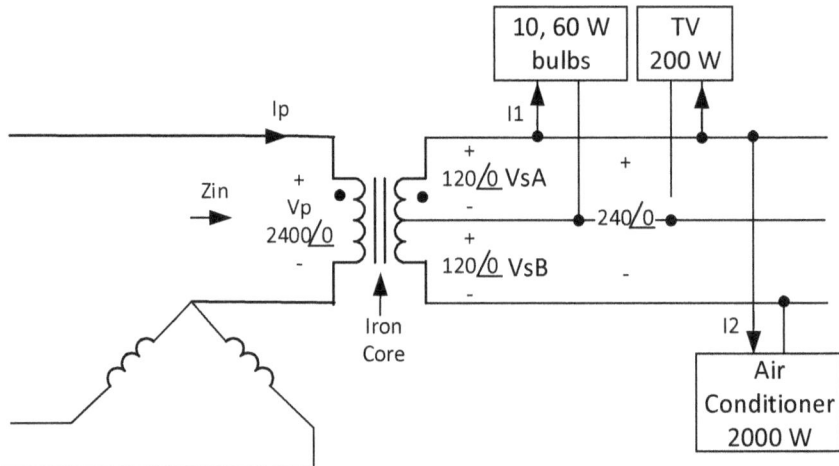

Fig. 5.3

```
%PTB2_Ex_82.mlx
c1=coeff(1/sqrt(3)); c3=coeff(3);    %some coefficients for use in calculations
P1=phasor(10*60, 0);                 %power in 10 bulbs
P2=phasor(2000, 0);                  %power in air conditioner
P3=phasor(200, 0);                   %power in TV
PT=P1+P2+P3                          %Total power in one phase
PT =
  phasor with properties:
    Mag: 2800
    phase: 0
Vp=phasor(2400, 0);
VsA=phasor(120, 0);
VsB=phasor(240, 0);
%part a)
Ip=PT/Vp                             %current in the primary side of the phase under consideration
Ip =
  phasor with properties:
    Mag: 1.1667
    phase: 0
Zin=Vp/Ip                            %effective resistance as seen in the primary side of the phase under consideration
Zin =
  phasor with properties:
    Mag: 2.0571e+03
    phase: 0
```

```
%part b)
I1=P1/VsA                  %current in all 10 bulbs
I2=P2/VsB                  %current in air conditioner
I2 =
  phasor with properties:
    Mag: 8.3333
    phase: 0
I1 =
  phasor with properties:
    Mag: 5
    phase: 0
%part c)
VLL=c1*Vp                  %line-line voltage
VLL =
  phasor with properties:
    Mag: 1.3856e+03
    phase: 0
%part d)
P3T=c3*PT                  %power in all 3 phases
P3T =
  phasor with properties:
    Mag: 8400
    phase: 0
```

MATLAB Ex_5.3 A transformer is used to match the 120 V source for maximum power transformer. The ac source has an internal source resistance of 500 ohms. Find the turns ratio of the transformer, the current I_P and the amount of power delivered by the ac source.

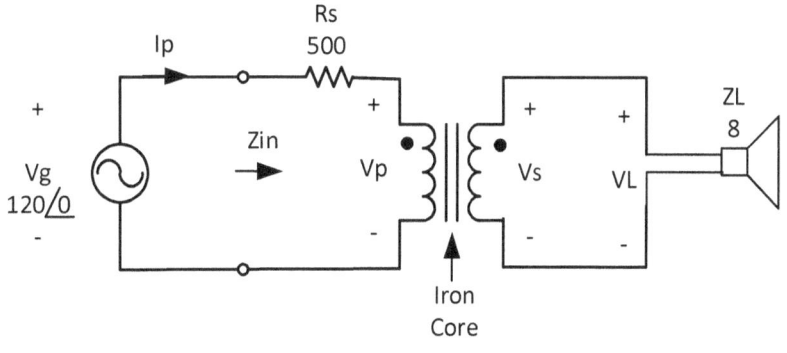

Fig. 5.4

```
%PTB_Ex_83.m
Rs=phasor(500, 0);    ZL=phasor(8, 0);
Zin=Rs;                    %under the maximum power transferred case
Vg=phasor(120, 0);
A=Rp/ZL;
a=sqrt(A.Mag)              %turn ratio is approx. 8
a = 7.9057
```

```
Ip=Vg/(Rs+Zin)
Ip =
  phasor with properties:
     Mag: 0.1200
    phase: 0
[S, P]=PWR(Vg, Ip, 'ph')
S =
  phasor with properties:
     Mag: 14.4000
    phase: 0
P =
  phasor with properties:
     Mag: 14.4000
    phase: 0
```

MATLAB Ex_5.4   Figure 5.4 a) shows a ac source V_g transformer equivalent circuit delivering power to load ZL. Figure 5.5 b) shows the circuit when the transformer is replaced by its equivalent impedances as seen on the primary winding side. The current on primary side is measured as $I_P = 10 \angle 0$. Find the source voltage V_g and the total impedance Z_{in} as seen by the source,

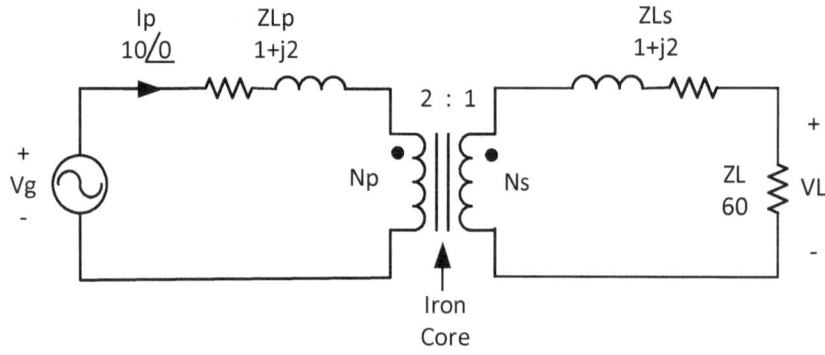

Fig. 5.5 a)

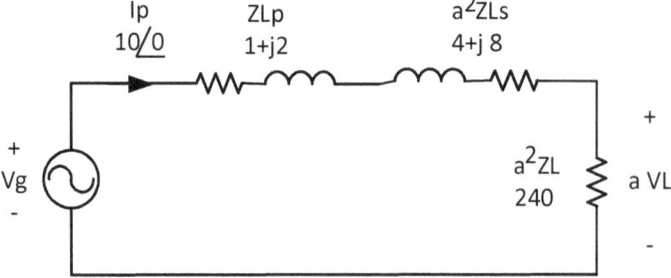

Fig. 5.5 b)

%PTB2_Ex_84.mlx
ZLp=x2ph(1+j*2, 'po'); ZLs=ZLp; a=coeff(2); a2=coeff(2^2);

```
Ip=phasor(10, 0);
ZL=phasor(60, 0);
Zin=ZLp+a2*ZLs;               %Load resistance reflected on the primary side
VL=Ip *a2* ZLs/a             %Load voltage phasor reflected on the primary side
VL =
  phasor with properties:
      Mag: 44.7214
    phase: 63.4349
Vg=Ip * Zin        %generator voltage phasor
Vg =
  phasor with properties:
      Mag: 111.8034
    phase: 63.4349
```

CHAPTER 6

3-PHASE CIRCUITS

LEARNING OBJECTIVES

- Phase to neutral and line to line voltage and Line currents in a generator
- Phase to neutral and line to line voltage and Line currents in a Y or Delta connected load.
- Draw phasor diagrams and time-domain waveforms.
- Power per phase and in all three phases
- Calculate per-phase and 3-phase Power Factor

CHAPTER INDEX

6.1   PHASOR TOOL BOX FUNCTIONS

PHASOR TOOL BOX FUNCTION: **[EAB, EBC, ECA] = phase2line(EAN, EBN, ECN)**
Converts phase to neutral voltage phasors to line-line voltage phasors.

PHASOR TOOL BOX FUNCTION: **[EAN, EBN, ECN] = line2phase(EAB, EBC, ECA)**
Converts line-line voltage phasors to phase to neutral voltage phasors.

PHASOR TOOL BOX FUNCTION: **[S, P, Q, Fp, ph]=PWR_3(Vph, Iph)**
Calculates the powers in 3-phase circuits. Phase to neutral voltages **Vph** and phase currents **Iph** are phasor arrays of 3 elements for 3 phases. **S** is apparent power, **Q** is reactive power and **P** is real power and **Fp** is the power factor. The **ph** indicates whether the power factor is lagging or leading type. **S, P** and **Q** are all polar quantities.

PHASOR TOOL BOX FUNCTION: **[S, P, Q, Fp, ph]=PWR_3line(VLL, IL)**
Calculates the powers in 3-phase circuits.
[S, P, Q, Fp, ph]=PWR_3(VLL, IL)   Line to line voltages **VLL** and Line currents **IL** are phasor arrays of 3 elements for 3 phases. **S** is apparent power, **Q** is reactive power and **P** is real power and **Fp** is the power factor. The **ph** indicates whether the power factor is lagging or leading type. **S, P** and **Q** are all polar quantities.

6.2   3-PHASE GENERATOR

MATLAB Ex_6.1   The 3-phase Y-connected generator is shown below. The generator supplies balanced load. The phase voltages and phase currents are as shown. The phase voltage is 120 ∠0 and the line current in phase A is 24 A ∠-53. Find the Line voltages, apparent, real and reactive power from phase A, the 3-phase total powers with the power factor.

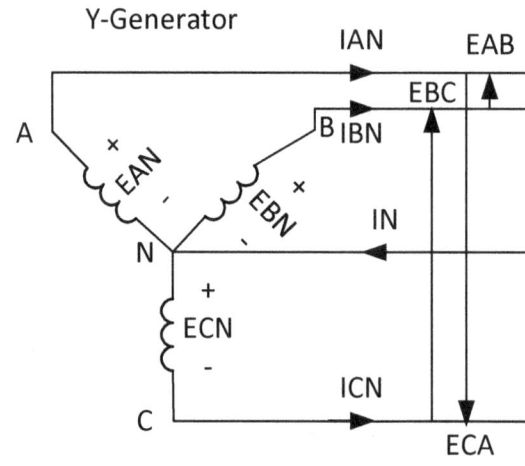

Fig. 6.1

```
%PTB2_Ex_115.mlx
%4-wire Y generator
clf
f=60; T=1/f;
EAN=phasor(120, 0);              %phase A voltage
EBN=phasor(120, -120)           %phase B voltage
EBN =
  phasor with properties:
    Mag: 120
   phase: -120
ECN=phasor(120, -240)           %phase C voltage
ECN =
  phasor with properties:
    Mag: 120
   phase: -240
Zan=x2ph(3+j*4, 'po');          %load impedance in phase A
Zbn=Zan;                         %load impedance in phase B
Zcn=Zan;                         %load impedance in phase C
EAB=EAN - EBN                    %Line A-Line B voltage
EAB =
  phasor with properties:
    Mag: 207.8461
   phase: 30.0000
EBC=EBN - ECN;                   %Line B-Line C voltage
```

```
ECA=ECN - EAN;              %Line C-Line A voltage
%alternately
c=coeff(sqrt(3));
EAB=c*EAN*rotate(30)        %multiply the phase voltage by sqrt(3) and rotate by 30 deg
EAB =
  phasor with properties:
    Mag: 207.8461
    phase: 30
%Alternately
[EAB, EBC, ECA]=phase2line(EAN, EBN, ECN)
EAB =
  phasor with properties:
    Mag: 207.8461
    phase: 30.0000
EBC =
  phasor with properties:
    Mag: 207.8461
    phase: -90
ECA =
  phasor with properties:
    Mag: 207.8461
    phase: 150
clf
phplot([EAN, EBN, ECN, EAB, EBC, ECA]); %phasor diagram
Current plot held
Current plot released
```

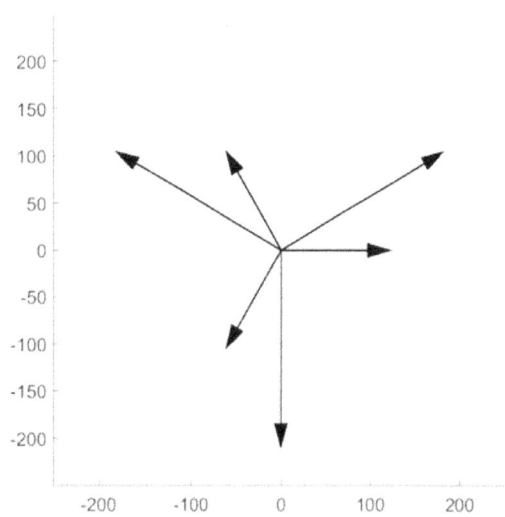

```
clf
phplot_signal([EAN, EBN, EAB, EBC], f, 0, T); %waveforms
```

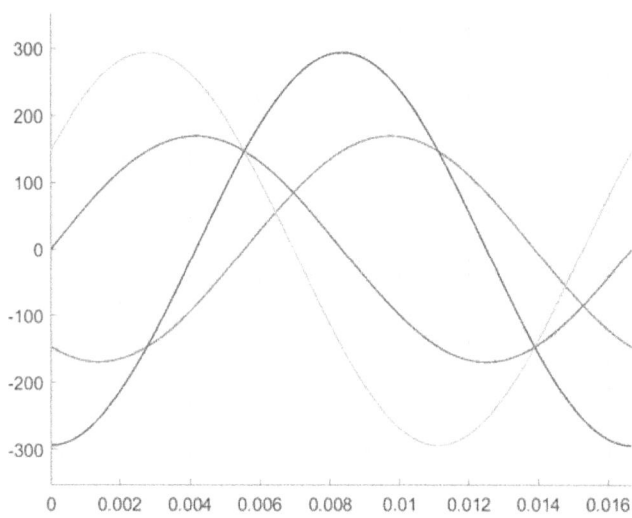

```
IAN=phasor(24, -53);        %Phase A current
IBN=IAN*rotate(-120);
ICN=IBN*rotate(-120);       %Phase C current
IN=IAN+IBN+ICN              %current in the neutral wire
IN =
  phasor with properties:
    Mag: 0
    phase: NaN
```

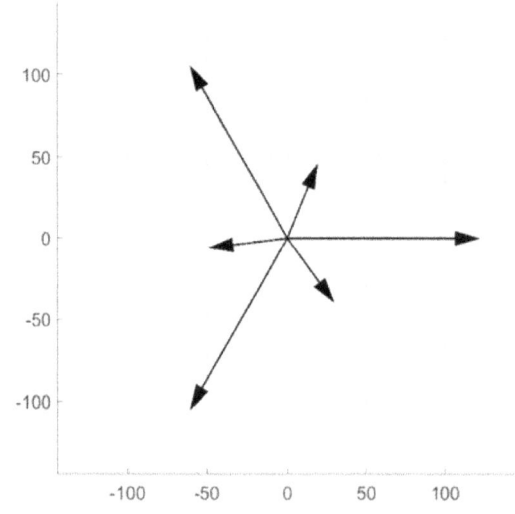

```
clf
c=coeff(2);
phplot([EAN, c*IAN, EBN, c*IBN, ECN,
c*ICN], 'ph');
```

```
clf
%phase current in phase A
[SA, PA, QA, FpA, ph]=PWR(EAN, IAN)
SA =
  phasor with properties:
```

```
    Mag: 2880
    phase: 53
PA =
  phasor with properties:
    Mag: 1.7332e+03
    phase: 0
QA =
  phasor with properties:
    Mag: 2.3001e+03
    phase: 90
FpA = 0.6018
ph = 'lagging'
```

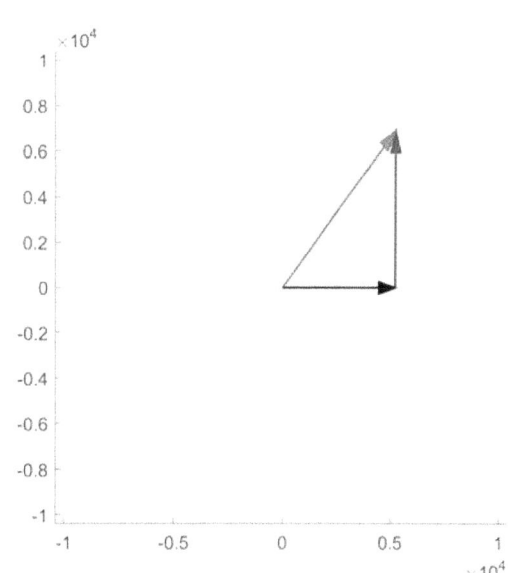

```
clf
c3=coeff(3);
P3=c3*PA
P3 =
  phasor with properties:
    Mag: 5.1997e+03
    phase: 0
Eph=[EAN, EBN, ECN];
Iph=[IAN, IBN, ICN];
%Alternately
ELL=[EAB, EBC, ECA];        %array of line voltages
IL=[IAN, IBN, ICN];         %Array of line currents
[S3, P3, Q3, Fp3, ph]=PWR_3line(ELL, IL)
S3 =
  phasor with properties:
    Mag: 8640
    phase: 53
P3 =
  phasor with properties:

    Mag: 5.1997e+03
    phase: 0
Q3 =
  phasor with properties:

    Mag: 6.9002e+03
    phase: 90
Fp3 = 0.6018
ph = 'lagging'
```

clf
add_graph(EAN, EBN, ECN)

ans =

  phasor with properties:

    Mag: 1.4211e-14

    phase: 0

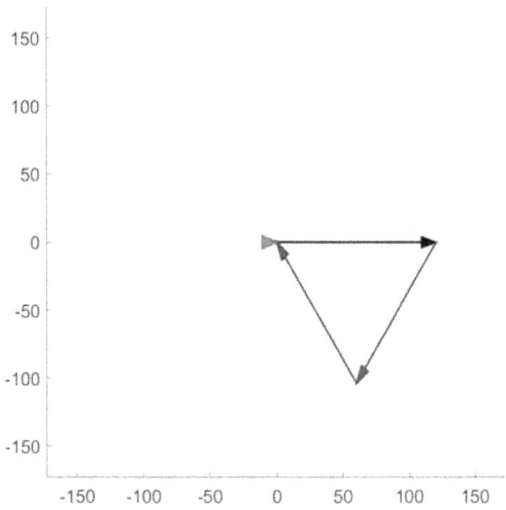

clf
add_graph(EAB, EBC, ECA)

ans =

  phasor with properties:

    Mag: 2.8422e-14

    phase: -90

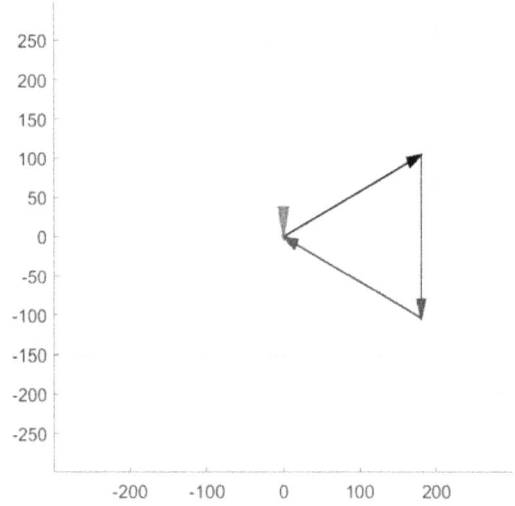

6.3   3-PHASE GENERATOR- LOAD CONFIGURATIONS

There are four possible configurations:

1) Y Generator – Y Load,   4-wire to 4-wire

2) Y Generator – Δ Load,   4-wire to 3-wire

3) Δ Generator – Y Load,   3-wire to 3-wire

4) Δ Generator – Δ Load, 3-wire to 3-wire

MATLAB Ex_6.2   The **4-wire, 3-phase Y-Y connected generator-load** is shown below. The generator phase voltages and the load impedances are as shown. Find the 3-phase total all 3 types of powers along with the power factor.

EAN=120∠0
EAN=120∠-120
EAN=120∠-240
Zan=Zbn=Zcn= 3+j 4

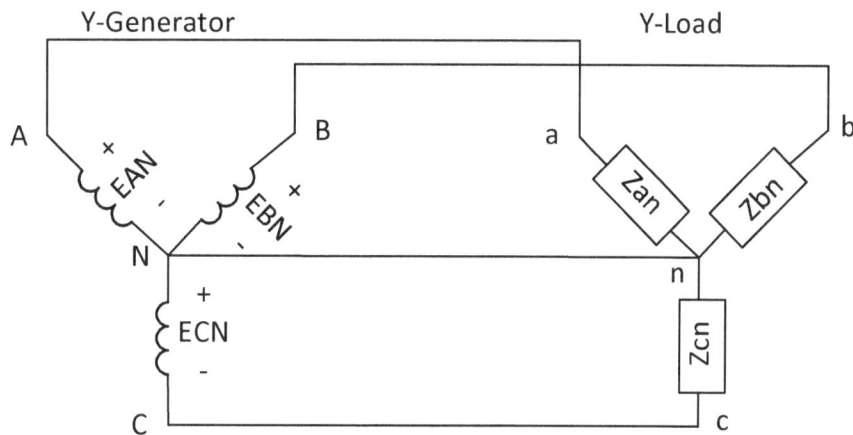

Fig. 6.2   Y-Generator with Y-Load

```
%PTB2_Ex_105.mlx
%4-wire Y-Y generator-load
clf
f=60; T=1/f;
EAN=phasor(120, 0);              phase A voltage
EBN=phasor(120, -120)           %phase B voltage
ECN=phasor(120, -240)           %phase C voltage
Zan=x2ph(3+j*4, 'po');          %load impedance in phase A
Zbn=Zan;                        %load impedance in phase B
Zcn=Zan;                        %load impedance in phase C
[EAB, EBC, ECA]=phase2line(EAN, EBN, ECN)        %Line voltages
Van=EAN;  Vbn=EBN;  Vcn=ECN; %load voltages in phase
Ian=Van/Zan                     %Phase A current
Ibn=Vbn/Zbn                     %Phase B current
Icn=Vcn/Zcn                     %Phase C current
In=Ian+Ibn+Icn                  %current in the neutral wire
In =
 phasor with properties:
  Mag: 1.2932e-14
```

phase: 164.0546
ELL=[EAB, EBC, ECA];   %array of line voltages
IL=[Ian, Ibn, Icn];   %Array of line currents
[S3, P3, Q3, Fp3, ph]=PWR_3line(ELL, IL)

S3 =
  phasor with properties:
     Mag: 8640
     phase: 53.1301
P3 =
  phasor with properties:
     Mag: 5.1840e+03
     phase: 0
Q3 =
  phasor with properties:
     Mag: 6.9120e+03
     phase: 90
Fp3 = 0.6000
ph = 'lagging'

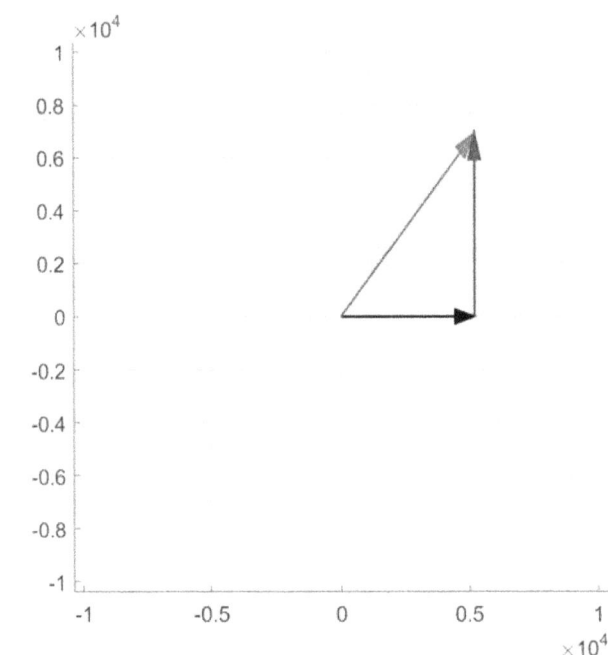

MATLAB Ex_6.3   The **3-wire balanced Δ- connected load** is shown below. The generator phase voltages and the load impedances are as shown. Find the 3-phase total all 3 types of powers along with the power factor.

EAB=150∠0
EBC=150∠-120
ECA=150∠-240
Zab=Zbc=Zca=6+j 8

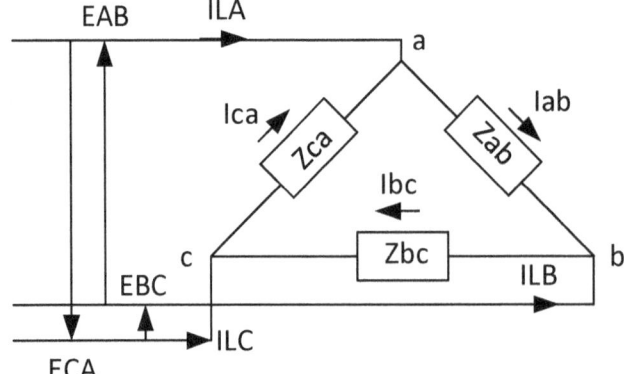

Fig. 6.3 3-wire balanced Δ- connected load

```
%PTB2_Ex_103.mlx
%3-wire Y-D, balanced load
clf
f=60; T=1/f;
EAB=phasor(150, 0);              %voltage between Lines A and B
EBC=phasor(150, -120);          %voltage between Lines B and C
ECA=phasor(150, -240);          %voltage between Lines C and A
Zab=x2ph(6+j*8, 'po');          %Balanced load impedance
Zbc=Zab;
Zca=Zab;
Iab=EAB / Zab                   %current between Lines A and B
Iab =
   phasor with properties:
      Mag: 21.2132
      phase: -53.1301
Ibc=EBC / Zbc;                  %current between Lines B and C
Ibc =
   phasor with properties:
      Mag: 21.2132
      phase: -173.1301
Ica=ECA/ Zca;                   %current between Lines C and A
Ica =
   phasor with properties:
      Mag: 21.2132
      phase: -293.1301
ILA=Iab-Ica                     %current in Line A
ILA =
   phasor with properties:
      Mag: 36.7423
      phase: -83.1301
clf
c=coeff(10);
phplot([EAB, c*Iab, c*Ica]);
Current plot held
Current plot released
clf
phplot_signal([EAB,ILA], f, 0, 2*T);   %plot of line voltage EAB and the line A current
```

Current plot held
Current plot released

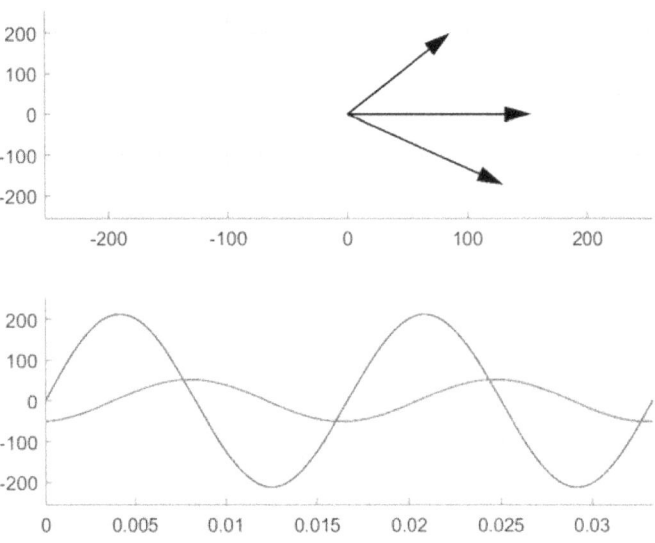

MATLAB Ex_6.4   The system with **3-wire balanced Δ- connected generated and Y-connected load** is shown below. Find the load and line voltages in all 3 phases, the total real power delivered by the generator and the power factor.

ILA=2 A∠0
Zan=Zbn=Zcn = 6- j 8 ohms

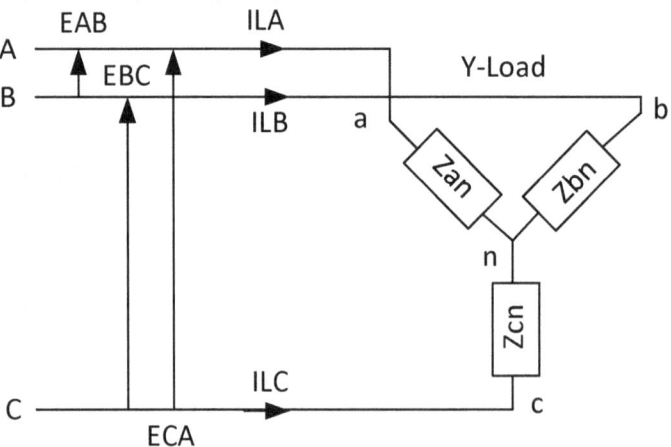

Fig. 6.4

```
clf, clear
ILA=phasor(2, 0);
ILB=ILA*rotate(120);
ILC=ILA*rotate(240);
Zan=x2ph(6-j*8, 'po');
Zbn=Zan;
Zcn=Zan;
```

Ean=ILA * Zan
Ean =
 phasor with properties:
    Mag: 20
    phase: -53.1301
Ebn=ILB * Zbn;
Ecn=ILC* Zcn;
 [EAB, EBC, ECA]=phase2line(Ean, Ebn, Ecn)
EAB =
 phasor with properties:
    Mag: 34.6410
    phase: -83.1301
EBC =
 phasor with properties:
    Mag: 34.6410
    phase: 36.8699
ECA =
 phasor with properties:
    Mag: 34.6410
    phase: 156.8699

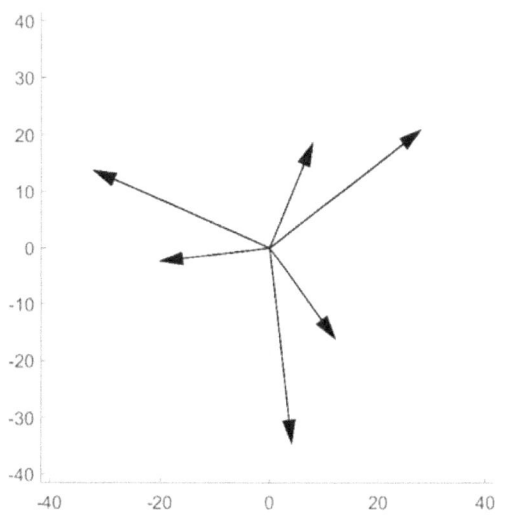

phplot([Ean, Ebn, Ecn, EAB, EBC, ECA]);
Current plot held
Current plot released
clf
V=[Ean, Ebn, Ecn];   I=[ILA, ILB, ILC];
[ST, PT, QT, Fp, ph]=PWR_3phase(V, I)
ST =
 phasor with properties:
    Mag: 120
    phase: -53.1301
PT =
 phasor with properties:
    Mag: 72
    phase: 0
QT =
 phasor with properties:
    Mag: 96
    phase: -90
Fp = 0.6000
ph = 'leading'

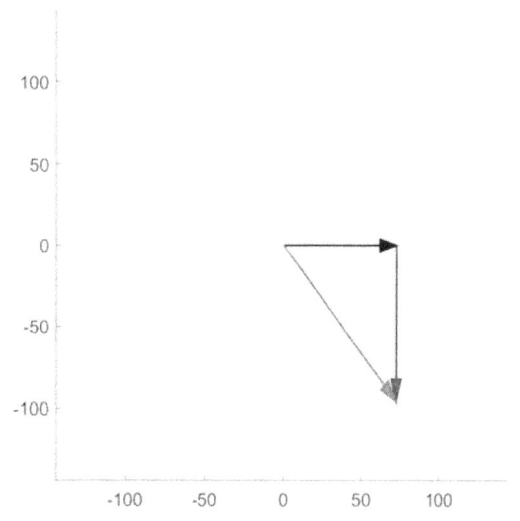

MATLAB Ex_6.5   The system with **3-wire balanced Y-Δ- connected load** is shown below. Find the load and line voltages in all 3 phases, the total real power delivered by the generator and the power factor.

EAB=200∠0
Zab=Zbc=Zca= 6-j 8 ohms
Zan=Zbn=Zcn= 4+j 3 ohms

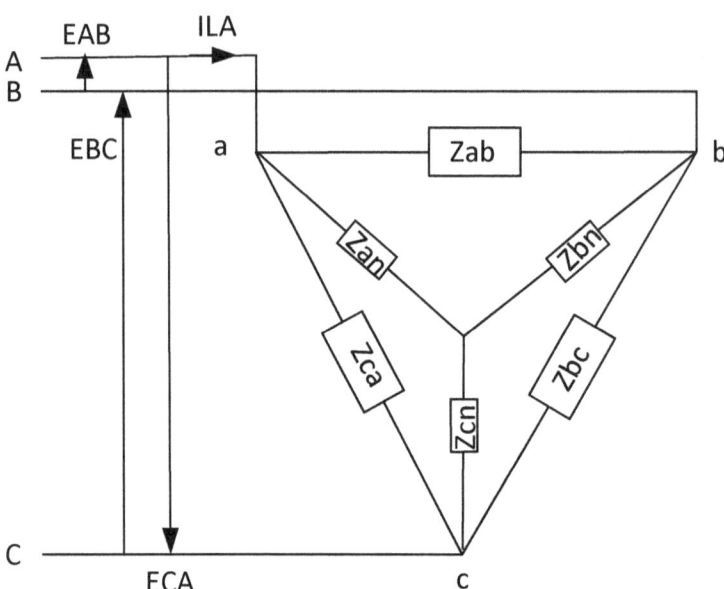

Fig. 6.5

```
clf
EAB=phasor(200, 0); EBC=phasor(200, -120); ECA=phasor(200, -240);
Zab=x2ph(6-j*8, 'po'); Zbc=Zab; Zca=Zab;          %Delta connected load
Zan=x2ph(4+j*3, 'po'); Zbn=Zan; Zcn=Zan;          %Y-connected load
%method1: first combine the Delta and Y loads then calculate powers.
%Convert delta-connected impedance load to Y-connected network
% combine the impedances in each branch
[Ean, Ebn, Ecn]=line2phase(EAB, EBC, ECA)
Ean =
  phasor with properties:
    Mag: 115.4701
   phase: 30
Ebn =
  phasor with properties:
    Mag: 115.4701
   phase: -90
Ecn =
  phasor with properties:
    Mag: 115.4701
   phase: -210
[Zan2, Zbn2, Zcn2]=delta2wye(Zab, Zbc, Zca)
Zan2 =
  phasor with properties:
```

    Mag: 3.3333
   phase: -53.1301
Zbn2 =
 phasor with properties:
    Mag: 3.3333
   phase: -53.1301
Zcn2 =
 phasor with properties:
    Mag: 3.3333
   phase: -53.1301
Zant=parallelZ([Zan,Zan2]);   Zbnt=parallelZ([Zbn,Zbn2]);   Zcnt=parallelZ([Zcn,Zcn2]);
Ian=Ean/Zant;   Ibn=Ebn/Zbnt;   Icn=Ecn/Zcnt;
Ed=[Ean, Ebn, Ecn];
Id=[Ian, Ibn, Icn];
[ST, PT, QT, Fp, phase]=PWR_3phase(Ed, Id)
ST =
 phasor with properties:
    Mag: 1.4422e+04
   phase: -19.4400
PT =
 phasor with properties:
    Mag: 1.3600e+04
   phase: 0
QT =
 phasor with properties:
    Mag: 4.8000e+03
   phase: -90
Fp = 0.9430
phase = 'leading'

%method 2 using Delta and Y loads
separately
%Delta Load
clf
Ed=[EAB, EBC, ECA];
Idab=EAB/Zab; Idbc=EBC/Zbc; Idca=ECA/Zca;
Id=[Idab, Idbc, Idca];
clf
[STd, PTd, QTd]=PWR_3phase(Ed, Id)
STd =
 phasor with properties:
    Mag: 12000
   phase: -53.1301
PTd =
 phasor with properties:
    Mag: 7200
   phase: 0

QTd =
  phasor with properties:
    Mag: 9600
    phase: -90

%Y-load
clf
[Van, Vbn, Vcn]=line2phase(EAB, EBC,ECA)
Van =
  phasor with properties:
    Mag: 115.4701
    phase: 30
Vbn =
  phasor with properties:
    Mag: 115.4701
    phase: -90
Vcn =
  phasor with properties:
    Mag: 115.4701
    phase: -210
V=[Van, Vbn, Vcn];
Ian=Van/Zan; Ibn=Vbn/Zbn; Icn=Vcn/Zcn;
I=[Ian, Ibn, Icn];
clf
[STy, PTy, QTy]=PWR_3phase(V, I)
STy =
  phasor with properties:
    Mag: 8.0000e+03
    phase: 36.8699
PTy =
  phasor with properties:
    Mag: 6400
    phase: 0
QTy =
  phasor with properties:
    Mag: 4.8000e+03
    phase: 90

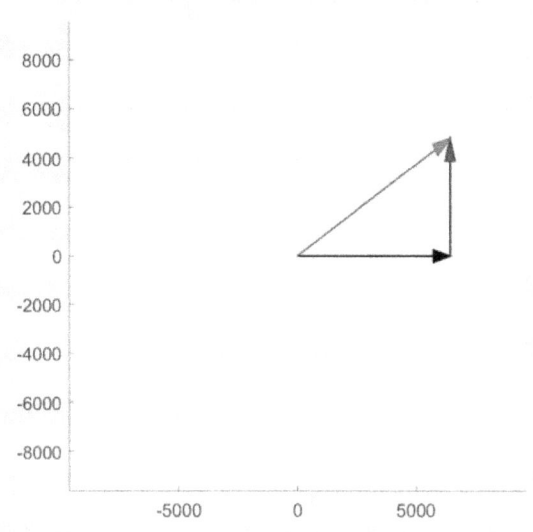

%full load calculations

PT=PTd+PTy

PT =

  phasor with properties:

    Mag: 13600

    phase: 0

QT=QTd+QTy

QT =

  phasor with properties:

    Mag: 4.8000e+03

    phase: -90

ST=STd+STy

ST =

  phasor with properties:

    Mag: 1.4422e+04

    phase: -19.4400

clf

triplot(ST)

Fp=PT.Mag /ST.Mag

Fp = 0.9430

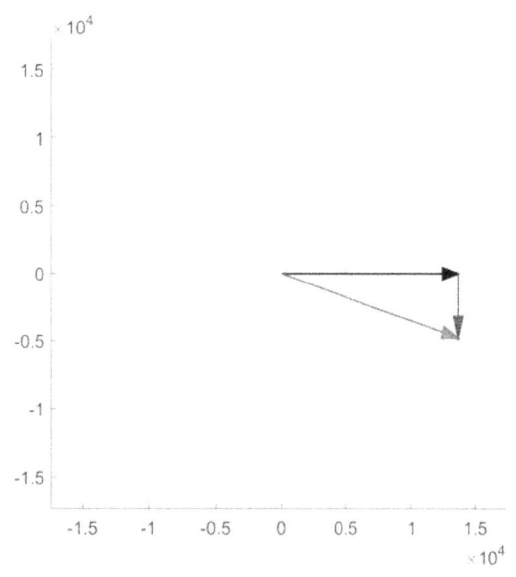

CHAPTER 7

PER UNIT ANALYSIS

LEARNING OBJECTIVES

- Understand the electrical ratings of electrical circuits
- Conversion from actual value to the per-unit values
- Conversion from the per-unit values to the actual values.

CHAPTER INDEX

7.1   PER UNIT SYSTEM

7.1.1   PER UNIT SYSTEM IN POWER CIRCUITS WITHOUT TRANSFORMERS

In the per-unit system, all phasor or non-phasor variables are normalized such that their nominal values are unity or less. There are two primary base values for normalizations:
1) Rated VA, denoted by S_B
2) Rated Voltage, denoted by V_B

These primary base values will remain constant at any point in a power circuit if the circuit does not use any transformers. The secondary base values at any point in the power circuit are derived from the primary base values:

Base value of current $\qquad\qquad I_B = \dfrac{S_B}{V_B}$

Base value of impedance $\qquad\quad Z_B = \dfrac{V_B^2}{S_B}$

Per unit values of actual S, V, I and Z at any point in the power circuit are calculated as

Per unit Apparent Power: $\qquad S_{pu} = \dfrac{S}{S_B}$

Per unit Voltage: $\qquad\qquad\quad V_{pu} = \dfrac{V}{V_B}$

Per unit Current: $\qquad\qquad\quad I_{pu} = \dfrac{I}{I_B}$

Per unit impedance: $\qquad\qquad Z_{pu} = \dfrac{Z}{Z_B}$

7.1.2   PER UNIT SYSTEM IN POWER CIRCUITS WITH TRANSFORMERS

A typical transformer, single-phase or 3-phase, has the high voltage (HV) side and the low voltage (LV) side. The VA ratings are constant across a transformer from the one side to the other, but the voltage and current ratings change across transformers. If we use the primary base values for normalization on one side (say the HV side), then the primary base values on the other side are calculated as the following:

Assume that the transformer turn ratio from HV side to LV side: 1/a,

Base value of VA on the LV side $S_{BLV} = S_{BHV}$

Base value of Voltage on the LV side $V_{BLV} = \dfrac{V_{BHV}}{a}$

Rest of the secondary values must be recalculated using new primary base values on the LV side of the transformer. Note that by making the above transformations, the per unit value of actual S, V, I and Z will not alter from HV side to the LV side. This property renders transformers irrelevant in ac circuit analysis. This is the primary advantage in using per unit values than the actual values.

After all calculations are done, the actual values of S, V, I and Z at any point in the circuit are calculated by:

$$S = S_{pu}S_B$$
$$V = V_{pu}V_B$$
$$I = I_{pu}I_B$$
$$Z = Z_{pu}Z_B$$

7.2   PHASOR TOOL BOX FUNCTIONS

PHASOR TOOL BOX FUNCTION: **Y = ph2pu(A, Sb, Vb, type)**
Converts phasor A to a per unit value. Sb is the rated apparent power, and Vb is the rated voltage. Both Sb and Vb are scalars. Y is the per unit phasor/polar
Y=ph2pu(A, Sb, Vb, 'V')   A is a voltage phasor
Y=ph2pu(A, Sb, Vb, 'I')   A is a current phasor
Y=ph2pu(A, Sb, Vb, 'Z')   A is an impedance polar
Y=ph2pu(A, Sb, Vb, 'S')   A is the apparent power polar

PHASOR TOOL BOX FUNCTION: **Y = pu2ph(A, Sb, Vb, type)**
Converts per unit phasor A to the actual phasor/polar value. Sb is the rated apparent power, and Vb is the rated voltage. Both Sb and Vb are scalars. Y is the actual phasor/polar value.
Y=pu2ph(A, Sb, Vb, 'V')   A is a voltage phasor
Y=pu2ph (A, Sb, Vb, 'I')   A is a current phasor
Y=pu2ph (A, Sb, Vb, 'Z')   A is an impedance polar
Y=pu2ph (A, Sb, Vb, 'S')   A is the apparent power polar

7.3   POWER CIRCUITS WITHOUT TRANSFORMERS

MATLAB 7.2.1: An incandescent bulb is rated for 120 Vrms and 500 watts. If the bulb is connected to an input voltage which is 20% higher than the rated voltage find, per unit current and actual current.

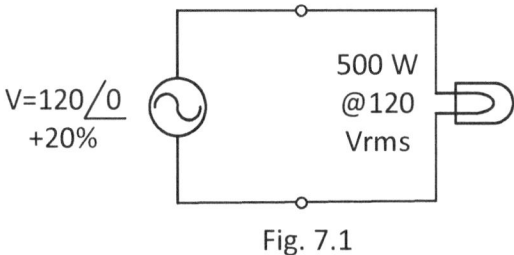

V=120$\underline{/0}$
+20%

500 W
@120
Vrms

Fig. 7.1

```
%PTB2_Ex_63A.m
Sb=phasor(500, 0);               %Rated VA
Vb=phasor(120, 0);               %Rated base voltage
Zpu=phasor(1.0, 0);              %per unit impedance
Vpu= ph2pu(V, 500, 120, 'V');    %convert actual voltage to per unit value
Ipu=Vpu / Zpu                    %per unit current
I= pu2ph(Ipu, 500, 120, 'I')     %actual current

Ipu =
 phasor with properties:
   Mag: 1.2000
  phase: 0
I =
 phasor with properties:
   Mag: 5
  phase: 0   phase: 0
```

7.4   POWER CIRCUITS WITH TRANSFORMERS

7.4.1   SINGLE-PHASE POWER CIRCUITS WITH TRANSFORMERS

MATLAB 7.3.1: A single-phase transformer rated 200 KVA, 200/400 V, and 10% short-circuit reactance. Compute the voltage regulation when the transformer is fully loaded at unity pf at the rated voltage 400 V.

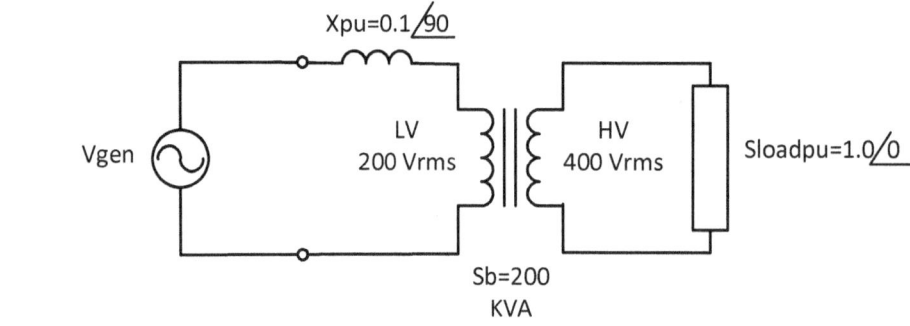

Fig. 7.2

```
%PTB2_Ex_64.m
clf
Sb=200e3;                          %KVA
Vb_HV=400;                         %HV side base values
Vb_LV=200;                         %LV side base values
%load side is HV side
Sloadpu=phasor(1, 0);              %per unit full load VA at unity pf
Vloadpu=phasor(1, 0);              %per unit load voltage
Iload=phasor(Sb/Vb_HV, 0)
Iloadpu=ph2pu(Iload, Sb, Vb_HV, 'I')     %full load
%LV-side calculations, source side is the LV-side
Xpu=x2ph(j*.1, 'cx')               %per unit leakage inductance
Vspu=Vloadpu+ Iloadpu * Xpu        %per unit full load VA at unity pf
Vs=pu2ph(Vspu, Sb, Vb_LV, 'V')     %actual source voltage
Vs_noloadpu=Vspu;                  %no load source voltage in pu
VR=100*(Vs_noloadpu.Mag-Vloadpu.Mag)/Vloadpu.Mag     %Voltage regulation at the source end
phplot([Vspu, Iloadpu, Vloadpu], 'ph')


Iload =
  phasor with properties:
    Mag: 500
   phase: 0
Iloadpu =
  phasor with properties:
    Mag: 1
   phase: 0
Xpu =
  phasor with properties:
    Mag: 0.1000
   phase: 90
```

Vspu =
    phasor with properties:
        Mag: 1.0050
        phase: 5.7106
Vs =
    phasor with properties:
        Mag: 200.9975
        phase: 5.7106
VR =
    0.4988

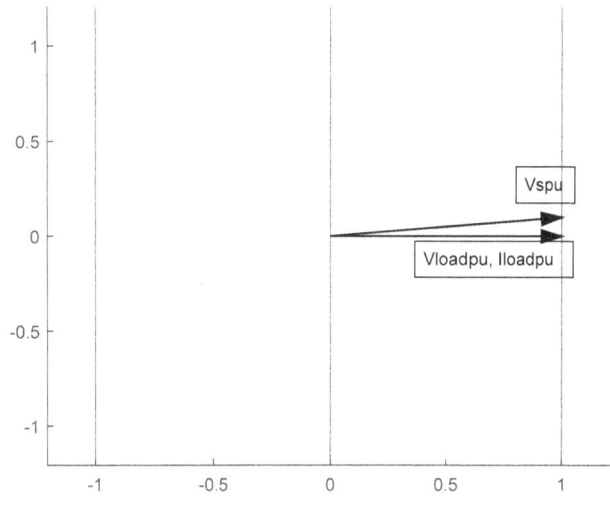

7.4.2   THREE-PHASE POWER CIRCUITS WITH TRANSFORMERS

MATLAB 7.3.2: A three-phase generator is supplying one MW load operating at 0.9 lagging pf to a grid. The generating station is 15 miles from the local power grid. The overhead line data for an aluminum steel conductor reinforced (ASCR) is given in appendix B.

a)   If the power is transferred at 460 V, select an overhead ASCR conductor to transfer the power. What is the generator voltage if the load voltage is specified at 460 Vrms AC?

b)   If the power is transferred at 3.3 kV, select an overhead ASCR conductor to transfer the power. What is the generator voltage if the load voltage is specified at 3.3 kVrms AC?

c)   If the power is transferred at 11.3 kV, select an overhead ASCR conductor to transfer the power. What is the generator voltage if the load voltage is specified at 3.3 kVrms AC?

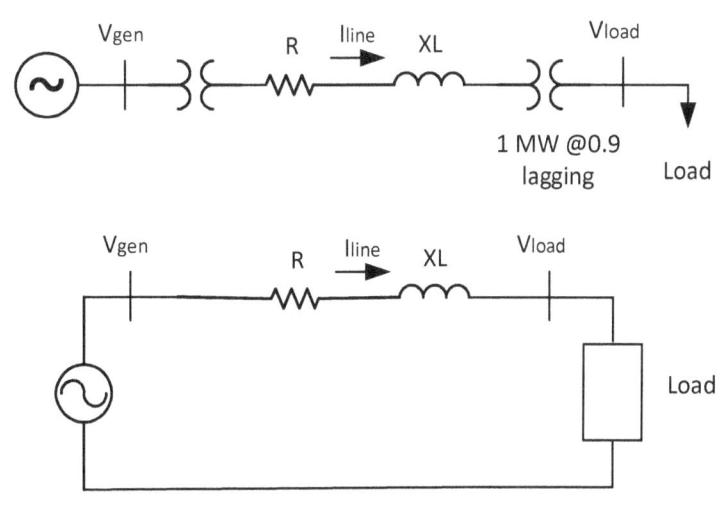

Fig. 7.3

```
%PTB2_Ex_6C5.mlx
%2_11, page 79
Sb=phasor(1e6,-acosd(0.9)) ;
c1=coeff(sqrt(3));
c2=coeff(3);
pf=coeff(0.9);
l=15;                    %Length of the transmission line
%a) VLL=460 Vrms
VLL=phasor(460, 0);
ILoad=Sb/(c1*VLL*pf)
ILoad =
  phasor with properties:
    Mag: 1.3946e+03
    phase: -25.8419
ro=0.0591; indreac=0.359; capreac=0.0814e6;
R=0.0591; XL=0.359;
Zline=x2ph((R+j*XL)*l, 'po');
VLoad_phi=VLL/c1
VLoad_phi =
  phasor with properties:
    Mag: 265.5811
    phase: 0
Vgen_phi=VLoad_phi+ILoad*Zline
Vgen_phi =
  phasor with properties:
    Mag: 7.7669e+03
    phase: 53.2083
VgenLL=c1*Vgen_phi
VgenLL =
  phasor with properties:
    Mag: 1.3453e+04
    phase: 53.2083
Sloss=c2*ILoad*conj(ILoad)*Zline
Sloss =
  phasor with properties:
    Mag: 3.1841e+07
    phase: 80.6516
PL=real(ph2x(Sloss, 'po'))
PL = 5.1722e+06
QL=imag(ph2x(Sloss, 'po'))
QL = 3.1418e+07
%b) VLL=3.3 KV Vrms
VLL=phasor(3300, 0);
ILoad=Sb/(c1*VLL*pf)
ILoad =
  phasor with properties:
    Mag: 194.3940
```

```
    phase: -25.8419
R=0.350; XL=0.465;
Zline=x2ph((R+j*XL)*I, 'po');
VLoad_phi=VLL/c1
VLoad_phi =
  phasor with properties:
    Mag: 1.9053e+03
    phase: 0
Vgen_phi=VLoad_phi+ILoad*Zline
Vgen_phi =
  phasor with properties:
    Mag: 3.5017e+03
    phase: 12.7941
VgenLL=c1*Vgen_phi
VgenLL =
  phasor with properties:
    Mag: 6.0652e+03
    phase: 12.7941
Sloss=c2*ILoad*conj(ILoad)*Zline
Sloss =
  phasor with properties:
    Mag: 9.8970e+05
    phase: 53.0317
PL=real(ph2x(Sloss, 'po'))
PL = 5.9518e+05
QL=imag(ph2x(Sloss, 'po'))
QL = 7.9074e+05
%c) VLL=11.3 KV Vrms
VLL=phasor(11300, 0);
ILoad=Sb/(c1*VLL*pf)
ILoad =
  phasor with properties:
    Mag: 56.7699
    phase: -25.8419
R=0.350; XL=0.465; XC=0.0537e6;
Zline=x2ph((R+j*XL)*I, 'po');
VLoad_phi=VLL/c1
VLoad_phi =
  phasor with properties:
    Mag: 6.5241e+03
    phase: 0
IC1=VLoad_phi/x2ph(j*XC/I, 'po')
IC1 =
  phasor with properties:
    Mag: 1.8224
    phase: -90
ILine=ILoad+IC1
```

ILine =

  phasor with properties:

    Mag: 57.5876

    phase: -27.4740

Vgen_phi=VLoad_phi+ILine*Zline

Vgen_phi =

  phasor with properties:

    Mag: 6.9810e+03

    phase: 1.7804

VgenLL=c1*Vgen_phi

VgenLL =

  phasor with properties:

    Mag: 1.2091e+04

    phase: 1.7804

Sloss=c2*ILoad*conj(ILoad)*Zline

Sloss =

  phasor with properties:

    Mag: 8.4406e+04

    phase: 53.0317

PL=real(ph2x(Sloss, 'po'))

PL = 5.0760e+04

QL=imag(ph2x(Sloss, 'po'))

QL = 6.7438e+04

Appendix A

MATLAB TUTORIAL

TOOL: MATLAB R 2017

Getting started with arithmetic calculations in the Command Window.
Launch MATLAB from the Desktop/Laptop. The following window should open.

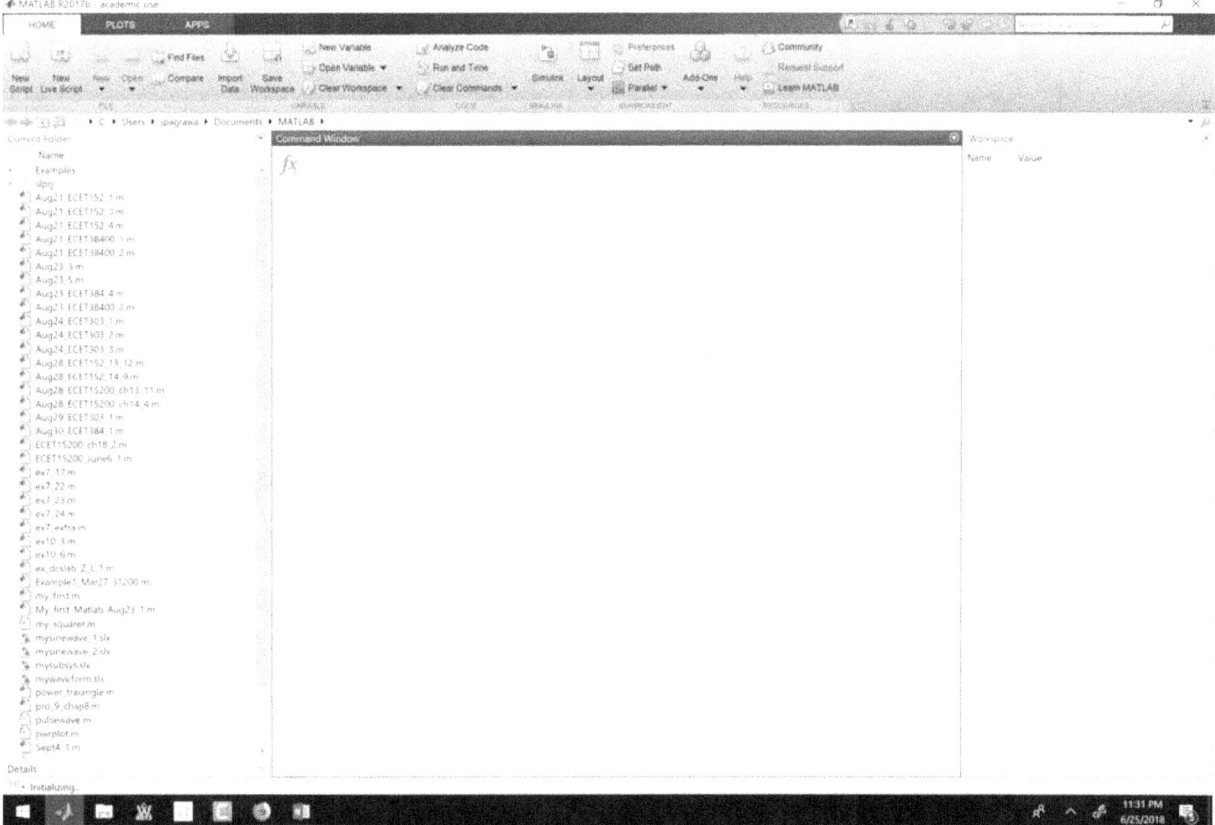

Below the Menu bar the default display has 4 major windows:
1) Path: from c drive to the current folder.
2) Current Folder: holds the current folder and the files and subfolders it contains.
3) Command Window: The main window where all executions are done and display the
                                    results whenever you want. It has prompt **>>** to enter an executable
                                                command.
4) Workspace: Area that displays all the variables and their latest values in the
                                                Command Window.

Demonstration in the Command Window
Open a new script file from the **New Script** in the menu bar. A new **Editor** window opens and the default display is as shown below:

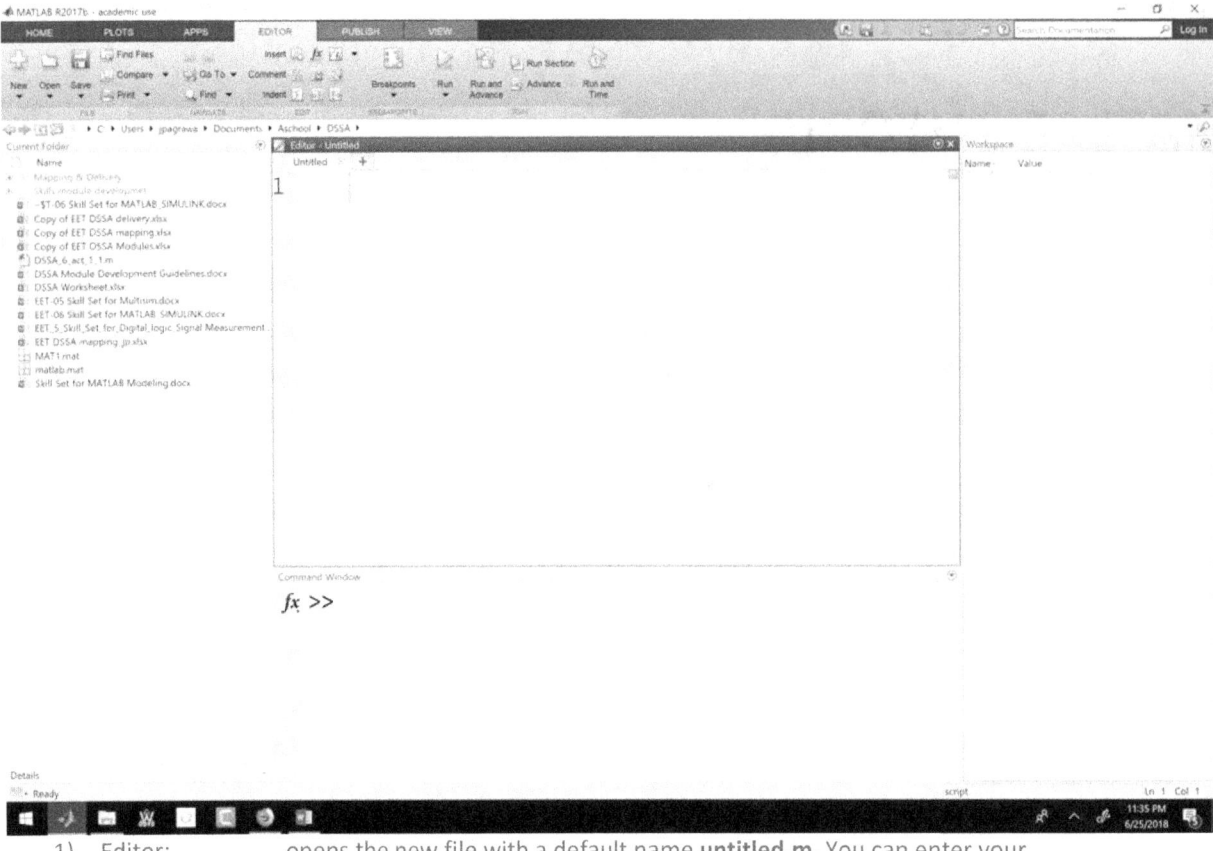

1) Editor: opens the new file with a default name **untitled.m.** You can enter your MATLAB code and save it under any valid file name whenever you wish. This window also contains files which you have selected to open or have opened so far in your current session.

Next, we will go ahead and perform some arithmetic calculations. Type in the command window as shown below and then press **enter**. Pressing **enter** executes the code entered on the line in the command window.
```
>> 4*5   %the code is 'multiply 4 by 5', execute the code and show the result on the next line, press enter now
ans =
   20
>>
```

Note that the MATLAB command is in black font. Anything written after % shows up in green font which indicates that it a non-executable statement, ignored by MATLAB but is very useful for the coder to understand the intent and reference of the line code in future. The result of the execution is shown in next lines, highlighted, The result is followed by the prompt **>>** indicating that the command window is ready for the next command. Observe that **ans** is the default variable where the result of the execution is stored. This default variable and its value is stored in the **Workspace** window.

Let us code more.

```
>> 4+2
ans =
   6
```

Note that the value of the variable **ans** in the workspace has modified to the new value 6.

```
>> 4+2-5
ans =
   1
```

```
>> 4+2*5          %execution is from left to right
ans =
   14
```

Note that the line comment starts with a %. Anything after % is ignored by MATLAB.

```
>> 4  +  2 *  5      %ignores blank spaces in the command.
ans =
   14
```

```
>> 4+2*5/6        %execution is from left to right
ans =
   5.6667
```

```
>> 4+2^3          %exponent has the highest priority, followed by * or /, + or -.
ans =
   12
```

```
>> (4+2)5/6
 (4+2)5/6
     ↑
Error: Unexpected MATLAB expression.
>>
```

Note that there is an error in writing the code, forgot to write '*' after the right parenthesis for multiplication operation. The Command window responds by a red remark showing error. The error display may not exactly tell what is the actual error. Therefore, examine your code line very seriously, eyeball it or ask for help. MATLAB does not presume anything. It needs exact syntax to act upon. Just retype the correct code on the next line at the prompt.

```
>> (4+2)*5/6      %parenthesis () have the highest priority in execution, again left to right
ans =
   5
```

```
>> x=(4+2)*5/6    %the right side of '=' is evaluated and the value assigned the variable x on
                        %the left of the '='.
x =
   5
```

Notice that the workspace has a new variable x with a value of 5.

```
>> x2= x*4+2      %the current value of variable x in the workspace is used for calculation.
x2 =
   22
```

Notice that the workspace has a new variable x2 with a value of 22. All old variables stay.

```
>> x3= x+x2           %the current values of variable x and x2 are used for calculation.
X3 =
   27
```

```
>> x4= x+x3;       %the command is executed but the result is not shown in the command
                      %window, however, the value of new variable x4 is visible in the
                      %workspace.
```

```
>> x4                 %this command asks for the value of the workspace variable x4.
x4 =
   32
```

```
>> save MAT_june_29   %saves all work in the command window in your current session
                          %into a file MAT_june_20.mat in the current folder, as seen in the
                          %left window.
```

```
>> clear              %clears all variables in the workspace window.
```

```
>> load MAT_june_29   %loads the contents and variables of the file in the workspace.
```

You can also load the file by double clicking on the file in the current folder window. Whereupon the command window shows the following command and the variables are loaded in the workspace.

```
>> load('MAT_june_29.mat')
```

You can copy all contents in a Word file by following steps:
1) Position the cursor in the command window.
2) Press **CTRL+A ,** which encloses all contents in the command window, shown in blue color.
3) Press **CTRL+C ,** to copy all contents in the command window.
4) Open a new WORD file on your computer, away from the MATLAB window.
5) Press **CTRL+V ,** to paste the contents that is copied in the previous step.
6) Save the file in your DSSA folder as MAT_june_29_1.doc.

It is the right time to introduce the priority in execution of mathematical operations on single element, be it a constant or a variable.

Priority of Elemental Operations:

Parenthesis	(expression)	all expressions within parenthesis are executed on the top property 0
Exponent	^	priority 1
Multiply/divide	* or /	priority 2
Add/subtract	+ or -	priority 3

In case of equal priority levels, the operations are executed from left to right.

Mathematics using Scripts

Demonstration:

The Command window programming is line by line execution and is completely interactive. Instead, we can also write all code lines in .m script and run them all in the given sequence by calling the .m file from the command window.

Launch MATLAB on your desktop. Go to editor and set the current folder to DSSA or wherever you wish to store your work.

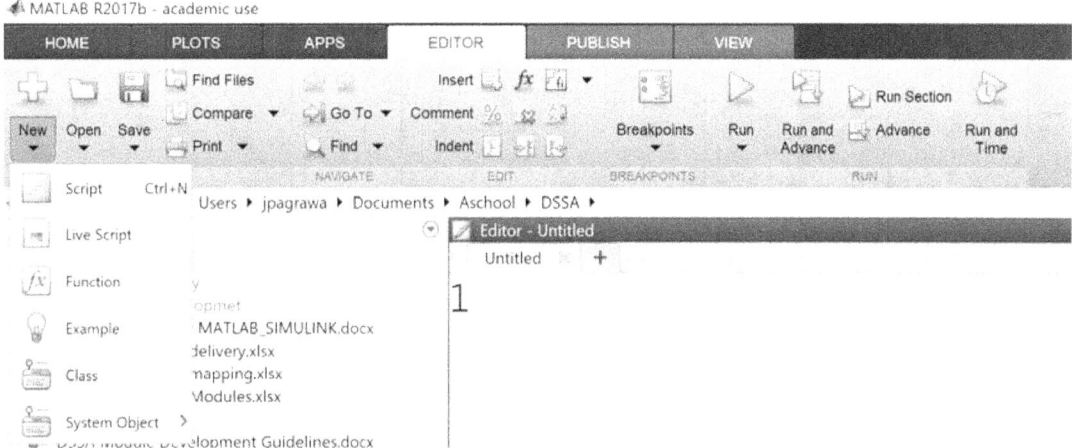

We will open a **New /Script.** The default name of the script is "Untitled". Enter the following code in the script. This code is same as what we did in the activity 1_1 in the command window.

```
%matlab_act1_3.m
4*5      %{ the code is 'multiply 4 by 5', execute the code and show the result on the next line, press enter now
```
Note that the MATLAB command is in black font. Anything written after % shows up in green font which indicates that it a non-executable statement, ignored by MATLAB but is very useful for the coder to understand the intent and reference of the line code in future. The result of the execution is shown in next lines, highlighted, The result is followed by the prompt >> indicating that the command window is ready for the next command. Observe that ans is the default variable where the result of the execution is stored. This default variable and its value is stored in the Workspace window.
```
%Let us code more. %}
4+2
```
%Note that the value of the variable ans in the workspace has modified to the new value 6.
```
4+2-5
4+2*5           %execution is from left to right
```
%Note that the line comment starts with a %. Anything after % is ignored by MATLAB.
```
4 + 2 * 5       %ignores blank spaces in the command.
4+2*5/6         %execution is from left to right
4+2^3           %exponent has the highest priority, followed by * or /, + or -.
(4+2)*5/6       %parenthesis () have the highest priority in execution, again left to right
x=(4+2)*5/6     %the right side of '=' is evaluated and the value assigned the variable x on
                %the left of the '='.
                %Notice that the workspace has a new variable x with a value of 5.
x2= x*4+2       %the current value of variable x in the workspace is used for calculation.
```
%Notice that the workspace has a new variable x2 with a value of 22. All old variables stay.
```
x3= x+x2        %the current values of variable x and x2 are used for calculation.
x4= x+x3;       %the command is executed but the result is not shown in the command
```

```
              %window, however, the value of new variable x4 is visible in the
              %workspace.
x4            %this command asks for the value of the workspace variable x4.
```

Save the script as **matlab_act1_3.m** in the current folder. Then enter the following in the command window to run the script:

>> matlab_act1_3

The results are displayed in the command window in the sequence they were executed from the .m script.

```
ans =
   20
ans =
    6
ans =
    1
ans =
   14
ans =
   14
ans =
    5.6667
ans =
   12
ans =
    5
x =
    5
x2 =
   22
x3 =
   27
x4 =
   32
```

Alternately, after entering the entire code in the "Untitled" script click the green arrow (**Run**) in the menu bar. MATLAB will prompt you to save it in script file whereupon you name it **matlab_act1_3**. The script will be saved and executed the code in the script line by line in the sequence from top to bottom.

You can copy all contents of the script and the command window in a Word file by following steps:

1) Position the cursor in the script area, the **Editor** window.
2) Press **CTRL+A ,** which encloses all contents in the editor window, shown in blue color.
3) Press **CTRL+C ,** to copy all contents in the editor window.
4) Open a new WORD file on your computer, away from the MATLAB window.
5) Press **CTRL+V,** to paste the contents that is copied in the previous step.
6) Change the position of the cursor in the command window.
7) Select the results in the command window and press **CTRL+C ,** to copy all contents in the command window.
8) Press **CTRL+V ,** to paste the contents that is copied in the previous step, after the contents which was pasted in step 5).

9) Save the entire file in your DSSA folder as matlab_act1_3.docx.
10) Print the above file.

Using Live Script
Demonstration:
Matlab programming using Live script is another method which is more interactive yet it involves batch processing like in the .m files. In this method, the results of executions are displayed in the editor window itself.
Launch MATLAB on your desktop. Go to editor and set the current folder to DSSA or wherever you wish to store your work.
We will open a **New /Live Script.** The default name of the script is "Untitled". Enter the following code in the script. This code is same as what we did in the activity 1_3 in the .m script.

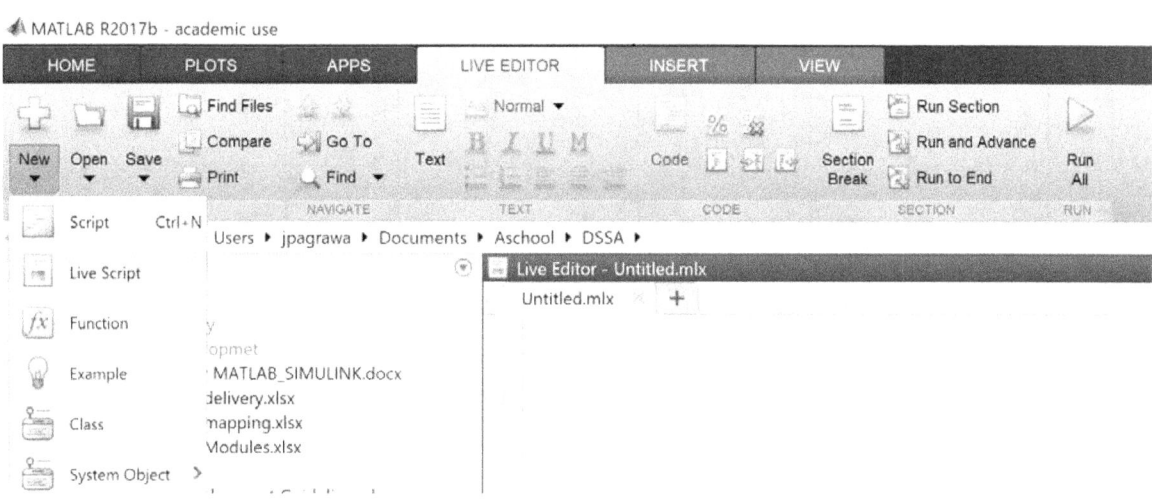

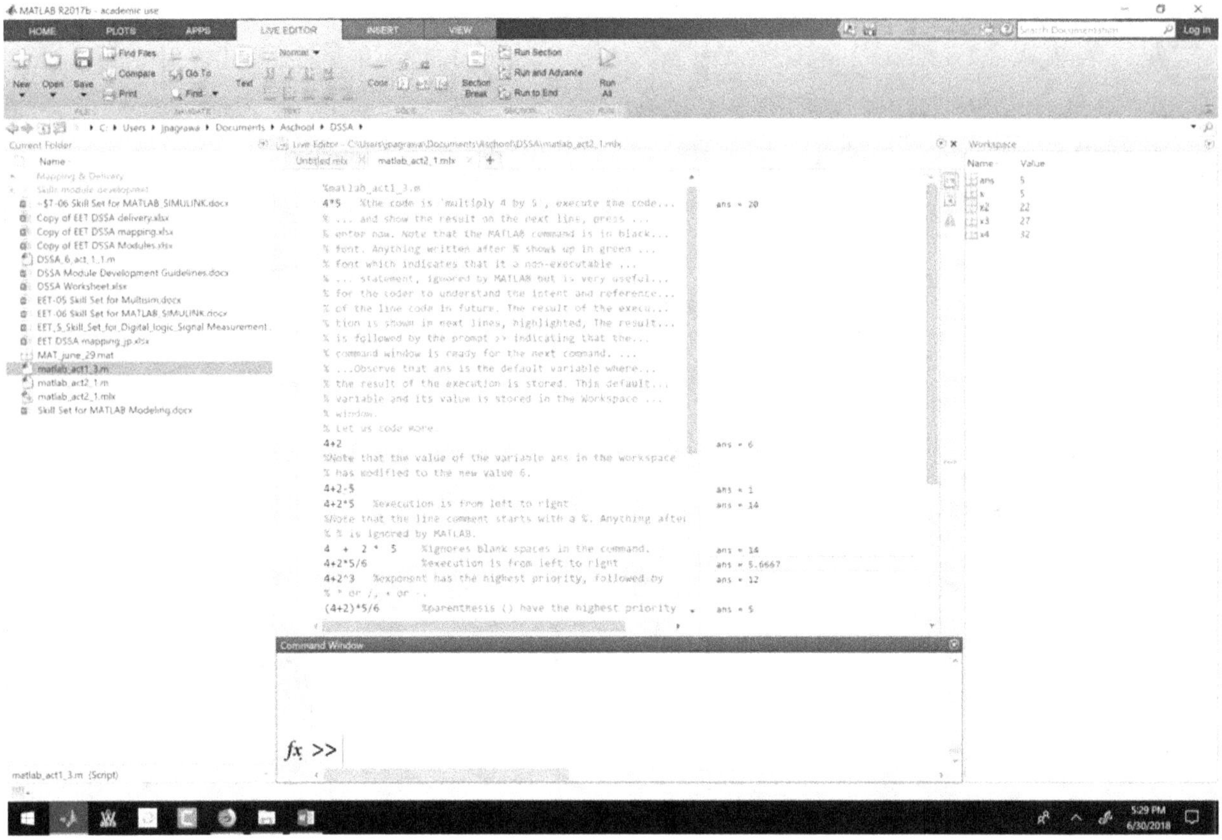

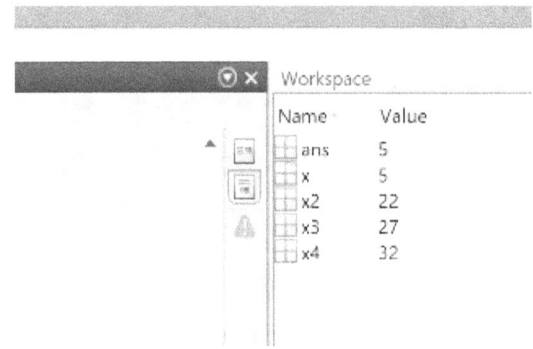

After entering the entire code in the "Untitled" script click the green arrow (**Run**) in the menu bar. MATLAB will prompt you to save it in script file whereupon you name it **matlab_act2_1** and select to save it as a **.mlx file**. The live script will be saved and executed line by line in the sequence from top to bottom. The results are shown to the right of that line.

Alternately, click the secind icon in the narrow window, to the **left of the Workspace** window as shown below:

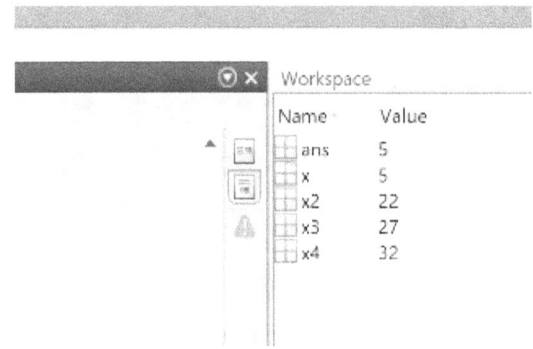

Results in the live script are now displayed below each line executed as shown below:

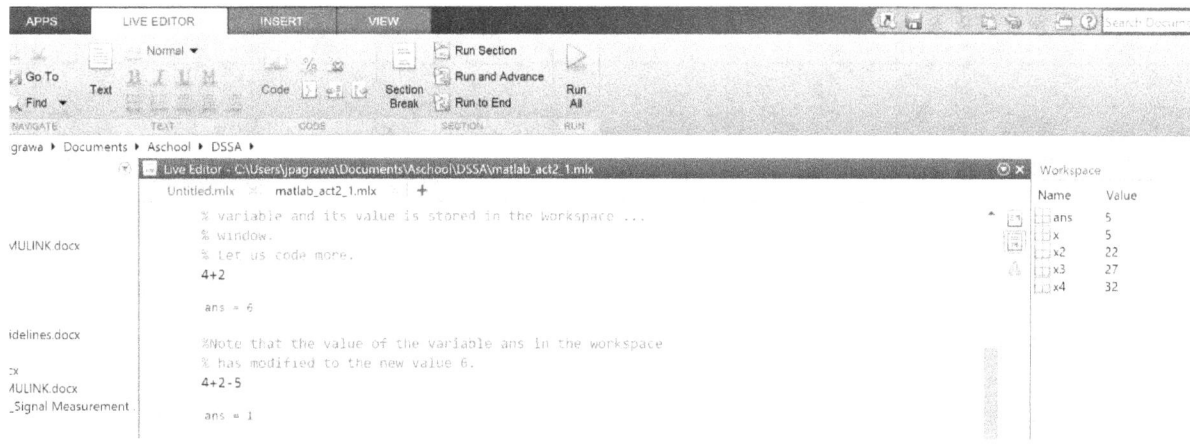

You can copy all contents of the script and the command window in a Word file by following steps:

1) Position the cursor in the script area, the **Editor** window.
2) Press **CTRL+A ,** which encloses all contents in the editor window, shown in blue color.
3) Press **CTRL+C ,** to copy all contents in the editor window.
4) Open a new WORD file on your computer, away from the MATLAB window.
5) Press **CTRL+V,** to paste the contents that is copied in the previous step.
6) Save the entire file in your DSSA folder as matlab_act2_1.docx.
7) Print the above file.

Array and Vector Calculations
Demonstration:
An Array is a horizontal sequence of values/ variables. A vector is a vertical array of values/ variables. The array elements are enclosed within square brackets.

x=[12, 5, 8, 10,.....]	Horizontal array of values
w=[x1, x2, x3, x4, ...]	Horizontal array of variables
d=[12	Vertical array (also called the vector) of values

```
        5
        8
        10
        .
        .
        . ]
```
A vector can also be written on a single line as

d=[12; 5; 8; 10; ...]

Problem statement:
Price of one apple is $0.50. Find the costs of options of buying 12, 5, 8 and 10 apples.
MATLAB Solution:

```
%EET06_act2_2.mlx
Price=0.50;              %elements separated by blank space or comma
Apples1=[12, 5, 8, 10]  %elements separated by blank space or comma
Apples =
   12   5   8   10
Cost=Apples1 * Price
```

Cost =
  6.0000   2.5000   4.0000   5.0000

%The arrays can also be organized as vectors, as shown below (continue the code):

Apples2=[12; 5; 8; 10] %element1s organized a vector

Apples2 =
   12
    5
    8
   10

Cost2=Apples2 * Price

Cost2 =
   6.0000
   2.5000
   4.0000
   5.0000

Elements of an n-element array A=[a_0, a_1, a_2, a_3, ...a_{n-1}] can be obtained as

   A(1)=a_0 A(2)=a_1 A(3)= a_2 A(n)=a_{n-1}

%EET06_act2_2B.mlx

x=[0, 1, 2, 3, 4, 5, 6, 7, 8, 9, 10]

x =
   0   1   2   3   4   5   6   7   8   9   10

x(1)   %first element of array x from left

ans = 0

x(end)   %last element of array x from left or the first element from the right

ans = 10

x(end-4)   %4th element from the right of array x

ans = 6

Generating uniformly spaced arrays:

Method 1

Syntax   x=initial value : increment or decrement : final value

To have or not to have the middle element is an option because the default increment is 1.

Example 1: x=0:1:10 will generate an array x=[0, 1, 2, 3, 4, 5, 6, 7, 8, 9, 10] with eleven elements.

Example 2: x=0:10 will generate the same array with a default increment of 1.

Example 3: x=0:-1:-10 will generate an array x=[0, -1, -2, -3, -4, -5, -6, -7, -8, -9, -10] with eleven elements.

%EET06_act2_2B.mlx

x1=0:10   %increasing array with default increment of 1
x1 =
   0   1   2   3   4   5   6   7   8   9   10
%N=length(x1)   %number of elements
x2=0:1:10   %increasing array
x2 =
   0   1   2   3   4   5   6   7   8   9   10
xb=0:-1:-10   %decreasing array
xb =
   0   -1   -2   -3   -4   -5   -6   -7   -8   -9   -10
xb2=-10:0   %increasing array
xb2 =
   -10   -9   -8   -7   -6   -5   -4   -3   -2   -1   0

Method 2

Syntax   x=linspace(initial value, final value, number of elements)

Example 1: x=linspace(0, 10, 11) will generate an array x=[0, 1, 2, 3, 4, 5, 6, 7, 8, 9, 10] with eleven elements.

x21=linspace(0, 10, 11)
x21 =
   0   1   2   3   4   5   6   7   8   9   10

Example 2: x=linspace(10, 0, 11) will generate an array x=[10, 9, 8, 7, 6, 5, 4, 3, 2, 1, 0] with eleven elements.

x22=linspace(10, 0, 11)
x22 =
   10   9   8   7   6   5   4   3   2   1   0

Array Manipulation

To convert a horizontal array to a vector and vice-versa

The transpose operator (') converts an horizontal array to a vertical array (vector) or vice versa. Another transpose operation will bring it back to the original array.

%Array manipulation
x1=x'   % convert the x array to a vector by an operator '
x1 =
    0
    1
    2
    3
    4
    5
    6
    7

```
   8
   9
  10
```
x2=x1'   % another transpose brings back the original x array
```
x2 =
   0   1   2   3   4   5   6   7   8   9   10
```

To make a sub-array out of an array

The operator ' : ' is a very useful operator to select a subsection of an array.

---

x5=x(1:5)   %a subarray comprising of elements 1 to 5 of x-array
```
x5 =
   0   1   2   3   4
```
x6=[x(1:3), x(6:8), x(end)] %a subarray comprising of elements 1 to 3, 6 to 8 and the last element of x-array
```
x6 =
   0   1   2   5   6   7   10
```

Horizontal Concatenation of Arrays

---

x7=[x5, x6]   %horizontal concatenation of array x5 and x6, obtained in the previous operations
```
x7 =
   0   1   2   3   4   0   1   2   5   6   7   10
```

Flip an array

```
x6
x6 =
   0   1   2   5   6   7   10
x8=flip(x6)
x8 =
  10   7   6   5   2   1   0
```

Sort an array

Syntax: sort(x, 'ascend') to sort the elements of array x in the ascending order from left.

        sort(x, 'descend') to sort the elements of array x in the descending order from left.

---

x9=sort(x7, 'ascend')
```
x9 =
```

```
0   0   1   1   2   2   3   4   5   6   7   10
```
x10=sort(x7, 'descend')
```
x10 =
   10   7   6   5   4   3   2   2   1   1   0   0
```

Maximum/ Minimum element of an array

Syntax: max(x)   to find the maximum element of the array x

min(x)   to find the minimum element of the array x

---

x11=max(x7)
```
x11 = 10
```
x12=min(x7)
```
x12 = 0
```

Sum/ cumulative Sum of elements of an array

Syntax: **sum(x)**   to find the sum of all elements of the array x

cumsum(x)   to find the cumulative sum of all elements to the left of that element including it in the

array x

---

x=[1, 3, 5, 8]
```
x =
   1   3   5   8
```
S=sum(x)
```
S = 17
```
S2=cumsum(x)
```
S2 =
   1   4   9   17
```

Product / cumulative Product of elements of an array

Syntax: **prod(x)**   to find the product of all elements of array x

cumprod(x)   to find the cumulative product of all elements to the left of that element including it in

the array x

---

x=[1, 3, 5, 8]
```
x =
   1   3   5   8
```
W=prod(x)

W = 120
W2=cumprod(x)
W2 =
   1   3 15 120

Find the number of elemnts in an array

Syntax: **length(x)**   to find the number of elements of an array x

---

x=[1, 3, 5, 8];
N=length(x)
N=
  4

Find an element in an array

Syntax: **n=find(x == value)**   to find the index of all elements that meet the criteria in array x. The criteria

involves relational operators like '==', '!=', '>', '<', '>=' or '<='.

---

x=[1, 3, 5, 8]
x =
   1   3   5   8
n=find(x==5)
n = 3
m=find(x>=3)
m =
   2   3   4

Mathematical operators are slightly different for arrays and vectors than single elements.

Priority of Array Operations:

Parenthesis	(expression)	all expressions within parenthesis are executed on the **top priority 0**
Exponent	.^	**priority 1**
Multiply/divide	.* or ./	**priority 2**
Add/subtract	+ or -	**priority 3**

In case of equal priority levels, the operations are executed from left to right. Furthermore, since an array can also have a single element, the priority levels 0, 1 and 2 can be used on single elements.

Problem statement:

Find and print the squares of all integers from 1 to 11.

MATLAB Code:_____

%EET06_act2_2B.mlx

x=0:10
x =

```
  0  1  2  3  4  5  6  7  8  9  10
y=x.^2              %squaring every element of x

y =
  0   1   4   9   16   25   36   49   64   81   100
```

```
%more array math
x1=[1, 3, 5, 8], x2=[0, -3, 6, 4]
x1 =
  1   3   5   8
x2 =
  0  -3   6   4
x3=x1.* x2      %elemental multiplication of x1 and the transpose of x2 arrays
x3 =
  0  -9   30   32
x3=[4, 3]
x3 =
  4   3
x4=x1 .* x3'  %elemental multiplication of x1 and the transpose of x3 arrays
x4 =
  4   12   20   32
  3   9   15   24

%evaluating a complex expression of arrays
w=2*x.^2+(x+3).*(x-1)
w =
  -3   2   13   30   53   82   117   158   205   258   317
```

Matrix Calculations
Demonstration:

A **mxn** Matrix has m rows and n columns, as shown below. The matrix elements are enclosed within square brackets.

$$X = \begin{bmatrix} 12 & 5 & 8 \\ -3 & 4 & -2 \end{bmatrix}$$

The MATLAB code for matrix on a single line

```
        X=[12, 5, 8 ; -3, 4, -2]
Or      X=[12, 5, 8
              -3, 4, -2 ]
```

It is customary to use uppercase letters for a Matrix variable. A popular use of matrix is in finding the solution of simultaneous equations.

Elements of an 3x3-matrix A given by

$$A = \begin{bmatrix} a0 & a1 & a2 \\ b0 & b1 & b2 \\ c0 & c1 & c2 \end{bmatrix}$$

can be obtained as

A(1, 1)=a0	A(1,2)=a1	A(1, 3)= a2
A(2, 1)=b0	A(2,2)=b1	A(2, 3)= b2
A(3, 1)=c0	A(3,2)=c1	A(3, 3)= c2

```
%EET06_act2_3.mlx
X1=[2, 4, 3; -1, 3, 6; 5, -2, 4]      %a 3x3 matrix
X1 =
   2    4    3
  -1    3    6
   5   -2    4
size(X1)    %size of matrix, the first number shows the #rows and the other the #columns
ans =
   3    3
X1(1, 1)
ans = 2
X1(1, end)
ans = 3
X1(2, end-1)
ans = 3
```

Matrix Manipulation

Transpose of Matrix

The transpose operator (') generates a new matrix which contains the columns of the original matrix as rows and the rows of the original matrix as the columns. Another transpose operation will bring it back to the original array.

```
XT=X1'               %transpose of a matrix
XT =
   2   -1    5
   4    3   -2
   3    6    4
Xn=XT'               %another transpose brings back the original matrix
Xn =
   2    4    3
  -1    3    6
   5   -2    4
```

Inverse of a Square Matrix

X^{-1} is the inverse of a square matrix (nxn) **X** . It is obtained by a standard MATLAB function.

Syntax: **inv(X)**   to find the inverse of a square matrix X

Note that X * X^{-1} = I where I is a unity matrix of size nxn.

$$I = \begin{bmatrix} 1 & 0 & 0 \\ 0 & 1 & 0 \\ 0 & 0 & 1 \end{bmatrix}$$

X1
X1 =
  2   4   3
 -1   3   6
  5   -2   4
X3=inv(X1)   %Inverse of a square matrix
X3 =
  0.1655 -0.1517   0.1034
  0.2345 -0.0483 -0.1034
 -0.0897   0.1655   0.0690
Xm=X1 * X3   %product of X and X^{-1} produces an Identity matrix
Xm =
  1.0000 -0.0000 -0.0000
 -0.0000   1.0000   0.0000
 -0.0000 -0.0000   1.0000

To make a sub-matrix out of a matrix

The operator ' : ' is a very useful operator to select a subsection of a matrix. If used in row, the ' : ' alone means all rows. If used in column, the ' : ' alone means all columns.

X1(1, 1)   %display the 1st row, first column element of the Matrix X1
ans = 2
X1(1, end)   %display the 1st row, last column element (also the first column from the right end) of ...
                                  the Matrix X1

ans = 3
X1(2, end-1)   %display the 2nd row, second column from the right end of the Matrix X1
ans = 3
X5=X1(1:2, 2:3)   %X5 contains the elements from rows 1 to 2 and columns 2 to 3 of X1 matrix
X5 =
  4   3
  3   6
X6=X1(1:2, :)   %X6 contains the elements from rows 1 to 2 and all columns of X1 matrix
X6 =
  2   4   3
 -1   3   6

Horizontal and Vertical Concatenation of Matrices

%Array manipulation

X7=[X5, X6] %horizontal concatenation of Matrices X5 and X6, both Matrixes must have equal …
number of rows.

X7 =

```
  4   3   2   4   3
  3   6  -1   3   6
```

X8=[X1; X6] %Vertical concatenation of Matrices X1 and X6, both Matrixes must have equal …
number of columns.

X8 =

```
  2   4   3
 -1   3   6
  5  -2   4
  2   4   3
 -1   3   6
```

Sort a Matrix

Syntax: **sort(X, 2, 'ascend')** to sort the row elements of matrix X in the ascending order from left to right.

'1' is for column-wise, '2' for row-wise.

sort(X, 2, 'descend') to sort the row elements of matrix X in the descending order from top to

bottom. 1' is for column-wise, '2' for row-wise.

X9=sort(X1, 2, 'ascend') %sort matrix X1 row wise in ascending order from left to right

X9 =

```
  2   3   4
 -1   3   6
 -2   4   5
```

X10=sort(X1, 1, 'descend') %sort matrix X1 column--ise in descending order from top to bottom

X10 =

```
  5   4   6
  2   3   4
 -1  -2   3
```

Maximum/ Minimum element in a Matrix

Syntax: **max(A, [], 2)** gives the row-wise maximum element in matrix A. '1' is for column-wise, '2' for row-wise.

min(A, [], 2) gives the row-wise minimum element in matrix A. '1' is for column-wise, '2' for row-wise.

X1

```
X1 =
   2    4    3
  -1    3    6
   5   -2    4
xmaxr=max(X1,[], 2)          %maximum element of each row in Matrix X1
xmaxr =
   4
   6
   5
xmaxc=max(X1, [],1 )         %maximum element of each column in Matrix X1
xmaxc =
   5    4    6
xminr=min(X1, [], 2)         %minimum element of each row in Matrix X1
xminr =
   2
  -1
  -2
```

Find the size of a matrix

Syntax: **size(X)**   to find the #rows and #columns of a matrix X

```
X1
X1 =
   2    4    3
  -1    3    6
[m, n]=size(X1)
m = 2
n = 3
```

Find an element in a matrix

Syntax: **n=find(x == value)**   to find the index of all elements that meet the criteria in matrix X. The criteria

involves relational operators like '==', '!=', '>', '<', '>=' or '<='.

```
[m, n]=find(X1==4)
m = 1
n = 2
```

Mathematical operators are slightly different for arrays and vectors than single elements.

Priority of Matrix Operations:

Parenthesis	(expression)	all expressions within parenthesis are executed on the **top priority 0**
Exponent	^	**priority 1**
Multiply/divide	* or /	**priority 2**
Add/subtract	+ or -	**priority 3**

In case of equal priority levels, the operations are executed from left to right and top to bottom.

X1
X1 =
```
   2    4    3
  -1    3    6
   5   -2    4
```
Y=X1^2   %every element of matrix X1 is squared
Y =
```
  15   14   42
  25   -7   39
  32    6   19
```
X1=[2, 4, 3; -1, 3, 6]   % a new matrix
X1 =
```
   2    4    3
  -1    3    6
```
X2=X1'
X2 =
```
   2   -1
   4    3
   3    6
```
X4=X1 * X2   % multiplication of mxn matrix to another and another nxk matrix ...
                % produces mxk matrix
x=[3; -1; 4]   % a vector
X21=X1 * x   %multiply a mxn matrix by a n-vector gives a mx1 vector
X4 =
```
  29   28
  28   46
```
x =
```
   3
  -1
   4
```
X21 =
```
  14
  18
```

Solving Equations

Objective: Learn to evaluate expressions involving one variable

Expressions with single variable:

Example 1:   Find the value of y in $y = 2x + 7$ when x=3.0. Verify using command window calculation.

Example 2:   Find the value of y in $y = 2x + 7$ when x=[3.0, 4.0, 5.5, 6.0]. Verify using command window calculation.

Example 3:   Find the value of y in $y = 2x^2 + 3x - 4$ when x=[3.0, 4.0, 5.5, 6.0]. Verify using command window calculation.

```
%EET06_act3_1.mlx
x=3.0
x = 3
y1=2*x+7                    % example 1
y1 = 13
x=[3.0, 4.0, 5.5, 6.0]
x =
    3.0000   4.0000   5.5000   6.0000
y2=2*x+7                    % example 2
y2 =
    13   15   18   19
y3=2*x.^2+3.*x-4           % example 3 note we use dot operators
y3 =
    23   40   73   86
```

More examples:

```
%EET06_act3_1.mlx
x=3.0
x = 3
y4=3*log10(200)+log(x)  %use of log10 and log(natural)
y4 =
    8.0017   8.2894   8.6078   8.6948
y5=4*exp(-2*x)+4      %use of exponential
y5 =
    4.0099   4.0013   4.0001   4.0000
th=pi/4;          %angle th=pi/4=45 deg
y6=3*sin(th)
y6 = 2.1213
y7=3*sin(2*th^2)+tan(3*th)
y7 = 1.8312
y8=3*sin(2*pi*x.^2)+tan(3*pi*x)  %notice that x is an array and dot operator is used
y8 =
    1.0e+15 *

    -0.0000  -0.0000   4.0932  -0.0000
```

Solving Equations of One Variable:
MATLAB can solve symbolic equations. This requires to declare symbolic variable and equations.
Syntax: syms var1 ... varN
symbolic variables var1 ... varN. Separate variables by spaces.
The equations are entered as the following example:
          eqn= 2*x^2_3*x+5==3.0
MATLAB has a function to solve the equation(s) by a **solve()**,
Syntax   **S = solve(eqn, var)**

solves the equation eqn for the variable var.

```
%EET06_act3_1B.mlx
syms x                    %declare x to be symbolic variable
eqn1=2*x+3==1.0;          %equation 1
solve(eqn1, x)            %x has a single value solution
ans = −1
eqn2=4*x^2+1==10          %equation 2 is a quadratic equation
```

$$eqn2 = 4\,x^2 + 1 = 10$$

```
solve(eqn2, x)            %x has two-value solution
ans =
```

$$\begin{pmatrix} -\dfrac{3}{2} \\[2ex] \dfrac{3}{2} \end{pmatrix}$$

```
eqn3=4*x^3-1==5;          %equation 3, since x has the highest power of 3, therefore, x will have 3 solutions
solve(eqn3, x)
ans =
```

$$\begin{pmatrix} \dfrac{2^{2/3}\,3^{1/3}}{2} \\[3ex] \dfrac{2^{2/3}\,3^{1/3}\left(-\dfrac{1}{2}+\dfrac{\sqrt{3}\,i}{2}\right)}{2} \\[3ex] -\dfrac{2^{2/3}\,3^{1/3}\left(\dfrac{1}{2}+\dfrac{\sqrt{3}\,i}{2}\right)}{2} \end{pmatrix}$$

```
eqn4 = sin(x) == 1;       %equation 4, an transcendental equation
solx = solve(eqn4, x)
solx =
```

$$\dfrac{\pi}{2}$$

```
eqn5 = sin(x^2) == 1;     %equation 5, x will have 2 solutions
solx = solve(eqn5, x)
solx =
```

$$\begin{pmatrix} -\dfrac{\sqrt{2}\,\sqrt{\pi}}{2} \\[3ex] \dfrac{\sqrt{2}\,\sqrt{\pi}}{2} \end{pmatrix}$$

Solving Simultaneous Equations

The solution for 2 variables in 2 simultaneous equations are obtained from,

eqn1=…….;
eqn2=…….;
S=solve([eqn1, eqn2], [x, y]);

The solutions **x and y** are obtained as **S.x** and **S.y**.

---

syms x y	%declare x and y as the symbolic variables
eqn1=x+y==0;	%equation 1
eqn2=x-2*y==1;	%equation 2
S=solve([eqn1, eqn2], [x, y]);	%the output S contains the solution for x and y
S.x	%solution for variable x

ans =

$$\frac{1}{3}$$

S.y	%solution for variable y

ans =

$$-\frac{1}{3}$$

Solving Simultaneous Equations using Matrix

Linear simultaneous equations in 2 variables are given in the following example:

$$x + y = 0$$
$$x - 2y = 1$$

When converted to matrix form,

$$\begin{bmatrix} 1 & 1 \\ 1 & -2 \end{bmatrix} \begin{bmatrix} x \\ y \end{bmatrix} = \begin{bmatrix} 0 \\ 1 \end{bmatrix}$$

A X = B

Where **A** is called the **Coefficient Matrix**, **B** is the **Source Vector** and **X** is the **Solution Vector**. We can either write the Matrix A and vectors B and X by visual inspection or via a standard MATLAB function, given below:

Syntax: [A, B] = equationsToMatrix(eqns, vars)

The solution for 2 variables in 2 simultaneous equations are obtained from,

X=A⁻¹B

Where **A⁻¹** is called the **inverse** of matrix **A**. **A⁻¹ B** is the matrix product of **inverse of A** and the vector **B**.

The MATLAB code for above is entered as the following:

X=A\B   %the '\' is called the back slash operator.

```
syms x y
[A, B] = equationsToMatrix([x + y == 0, x - 2*y == 1])
A =
```

$$\begin{pmatrix} 1 & 1 \\ 1 & -2 \end{pmatrix}$$

B =

$$\begin{pmatrix} 0 \\ 1 \end{pmatrix}$$

X=A\B

X =

$$\begin{pmatrix} \dfrac{1}{3} \\ -\dfrac{1}{3} \end{pmatrix}$$

Graphing

Objectives:

1) To plot a 2-dimensional graph.
2) Edit lines, axes, graphs, add labels, grid, colors etc.
3) Add multiple graphs on a single plot
4) Multiple plots on a single page
5) Linear, log and other types of graphs.

Demonstration

We will graph the relationship between two variables x and y given by the following table:

x	y
0	0
1	1
2	4
3	9
4	16
5	25
6	36
7	49
8	64
9	81
10	100

```
%EET06_act3_3.m
x=[0, 1, 2, 3, 4, 5, 6, 7, 8, 9, 10];              %x-array
y=[0, 1, 4, 9, 16, 25, 36, 49, 64, 81, 100];      %y-array
plot(x, y)                                         %the horizontal variable must be first.
grid                                               %add grid
title('y vs x graph')                              %add title
ylabel('y')                                        %add y-label
xlabel('x')                                        %add x-label
```

When run by pressing the green arrow on the menu bar, a graph appears in a separate window as shown below:

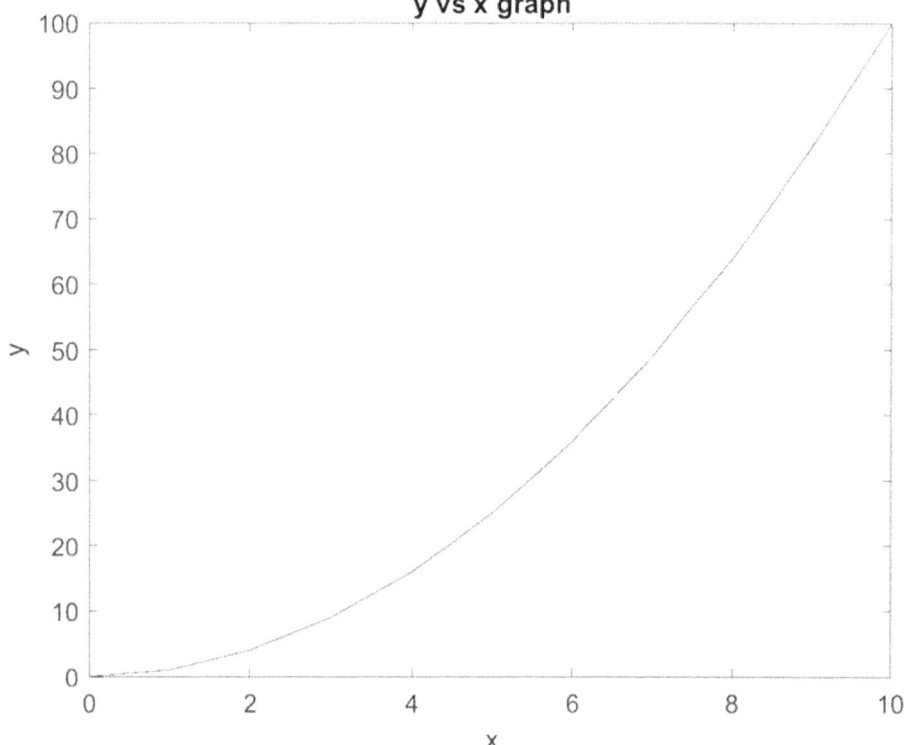

You can copy the figure by selecting **Figure 1/Edit/copy figure** and paste it in your Word file.

You can also add the grid, title, y-label and x-label on the **Figure 1** window directly by using the menu bar.

Open the **Insert** menu. There are various options. Let us add the title in the graph to a new title 'x-y graph'. Select the **title** from the dropped down menu and bring the mouse pointer in the graph window. A placement bar for entering text opens on the top of the graph, you can re-write the existing title or add one. After entering the text, Click the pointer arrow on anywhere on the graph, will terminate the process.

Next we will add a legend by selecting it from the dropped down menu and bring the mouse pointer in the graph window. The pointer arrow will change to a '+ 'sign. Move around the '+' sign and place at the desired point and click the left mouse button. A text bar will open wherein you type your text and then left click the pointer mouse anywhere in the graph area to terminate the process. Play with inserting other options yourself.

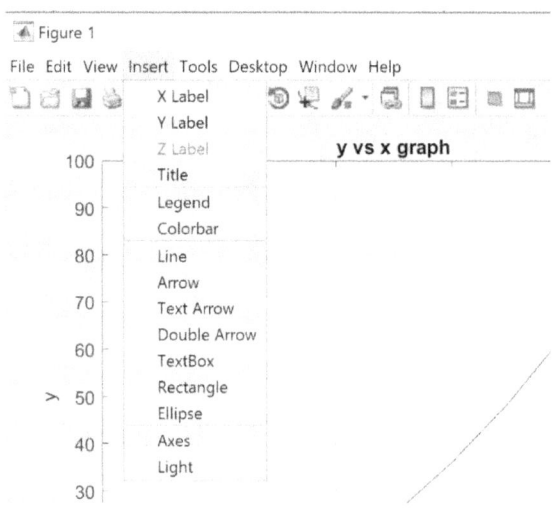

Formatting the Figure

We can edit the axes and colors and many other attributes by selecting the **Axes Properties** on the dropped down **Edit** menu. The existing x-axis limits are shown by default. For example, enter new values as -2 to 12 in the designated areas and press enter to effect the change. You can change other properties by clicking on '**More Properties...**'.

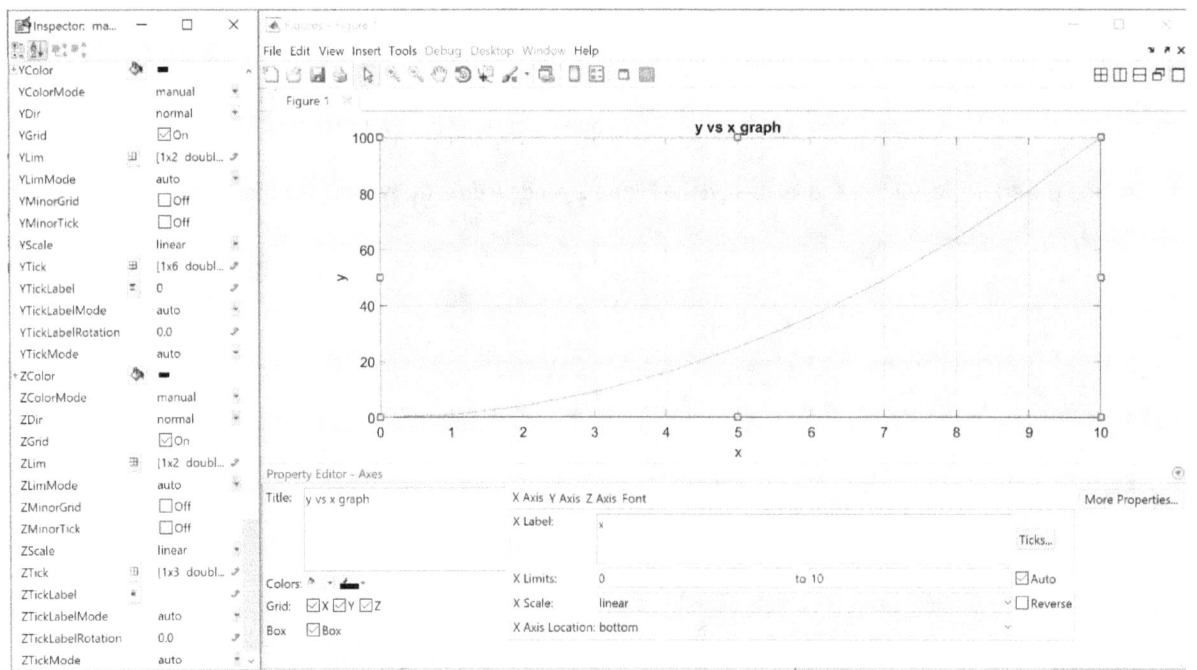

Another method is to get documentation help in MATLAB. Enter 'plot' in the 'search window' on the top right area of the main display window of the MATLAB. The documentation of all information on the plot function appears in the editor window.

For example, we will increase the thickness of the curve in the graph by modifying the plot command as the following:

plot(x, y, Linewidth, 3)

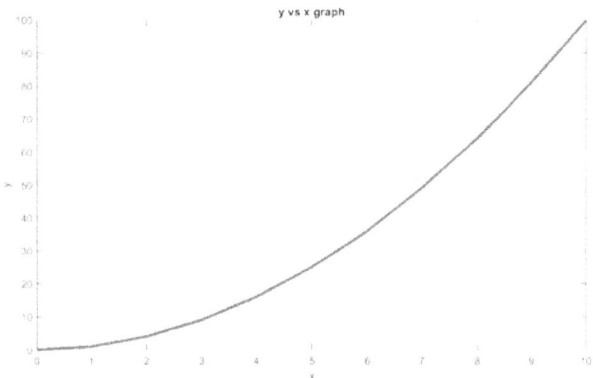

A graph or previous graphs can be cleared by entering **clf** on the command line or in the .m file or .mlx file.

Multiple Graphs on a Single Figure

Method 1:
Multiple plots on a graph can be added as the following code:
        plot(x1, y1, x2, y2, x3, y3,...)

The x1-y1, x2-y2, x3-y3 are the different curves. The syntax of the code should be maintained as shown here, given in the documentation as well or as. The overall graph has x-axis scaled to the maximum of x1, x2, x3... and the y-axis to the maximum of y1, y2, y3, ... An example is show below:

```
%EET06_act3_3A.m
clf
x1=0:10;              %x1-array
y1=x.^2;              %y1-array
x2=0:12;              %x2-array
y2=sqrt(100*x2);      %y2-array
%plot(x, y)
plot(x1, y1, 'red',  x2, y2, 'blue', 'linewidth', 3)
```

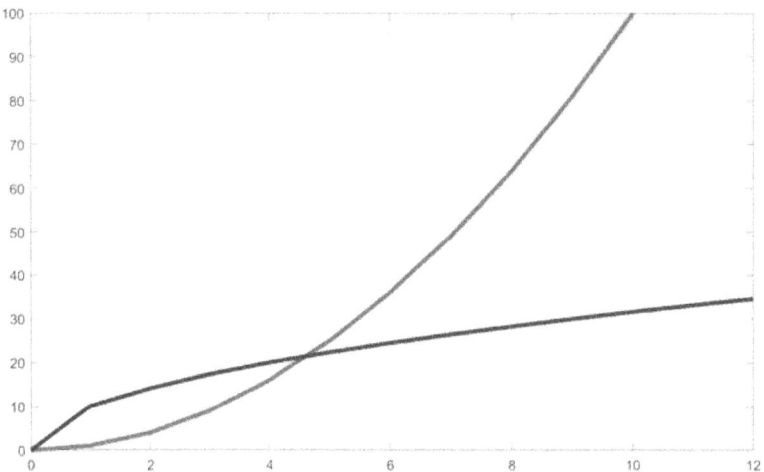

Method 2:

```
%EET06_act3_3A.m
clf
x1=0:10;              %x1-array
y1=x.^2;              %y1-array
x2=0:12;              %x2-array
y2=sqrt(100*x2);      %y2-array
%plot(x1, y1, 'red',  x2, y2, 'blue', 'linewidth', 3)
plot(x1, y1, 'linewidth', 1)
hold                  %hold the current graph and axes
plot(x2, y2, 'linewidth', 3)
hold                  %release the hold
```

```
>> EET06_act3_3_3A
Current plot held
Current plot released
```

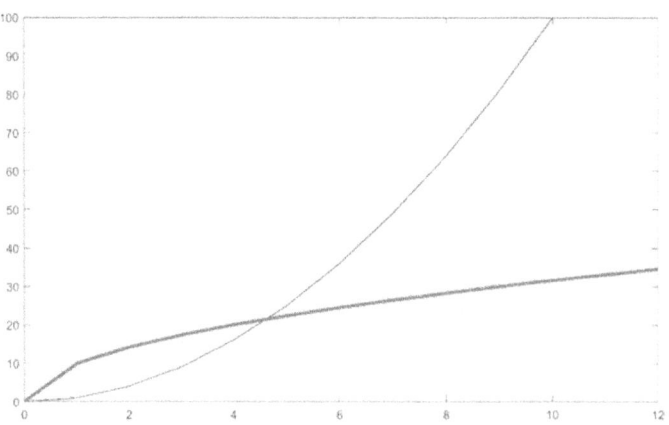

Multiple Figures on a Single Page

A single page is divided in matrix to obtain multiple figures. A 2:1 division is shown by an example:

MATLAB m_file: appendix_M _4.m_____

```
%appendix_M_4.m
x1=0:1:9;          y1=x1;
x2= 1: 1: 15;      y2=2 * x2;
```
% vertical organization
```
subplot(211)   %first plot on the top in a 2-row 1-column display
plot(x1, y1)
subplot(212)   %second plot on the bottom in a 2-row 1-column display
plot(x2, y2)
```

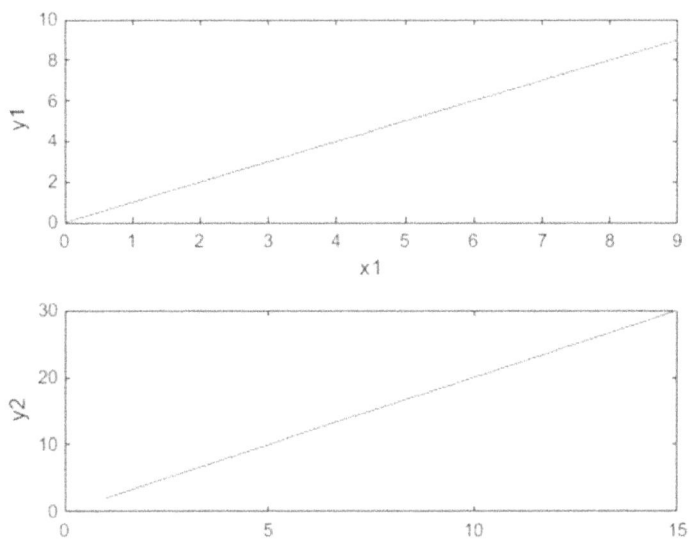

A 1:2 division is shown by an example:
% horizontal organization
```
subplot(121)              %first plot on the left in a 1-row 2-column display
plot(x1, y1)
subplot(122)              %second plot on the right in a 1-row 2-column display
plot(x2, y2)
```

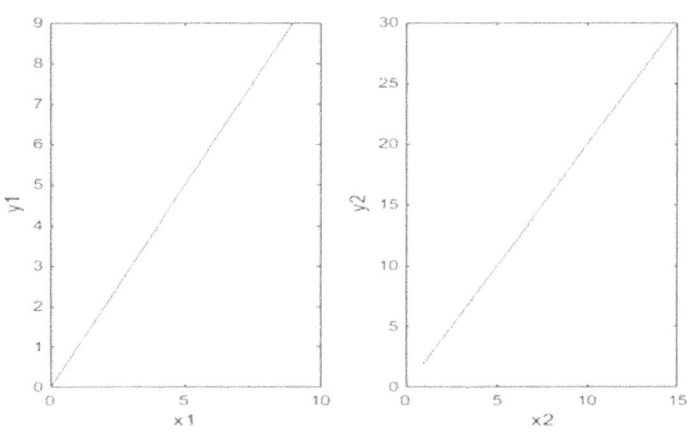

A 2:2 division is shown by an example:
% Matrix organization

```
%EET06_act3_3A.m
clf
x1=0:10;              %x1-array
y1=x.^2;              %y1-array
x2=0:12;              %x2-array
y2=sqrt(100*x2);      %y2-array
x3=2:12;              %x3-array
y3=sqrt(x3);          %y3-array
subplot(321)
plot(x1, y1, 'linewidth', 1)
subplot(322)
plot(x2, y2, 'linewidth', 3)
subplot(323)
plot(x3, y3)
```

Reading from the graph

We will read the minimum point on the curve as given below. The code for the graph is shown as following:

```
%EET06_act3_3C.m
clf
x=-5:.1:5;            %x-array
y=4*x.^2+5;           %y-array
plot(x, y, 'linewidth', 1)
```

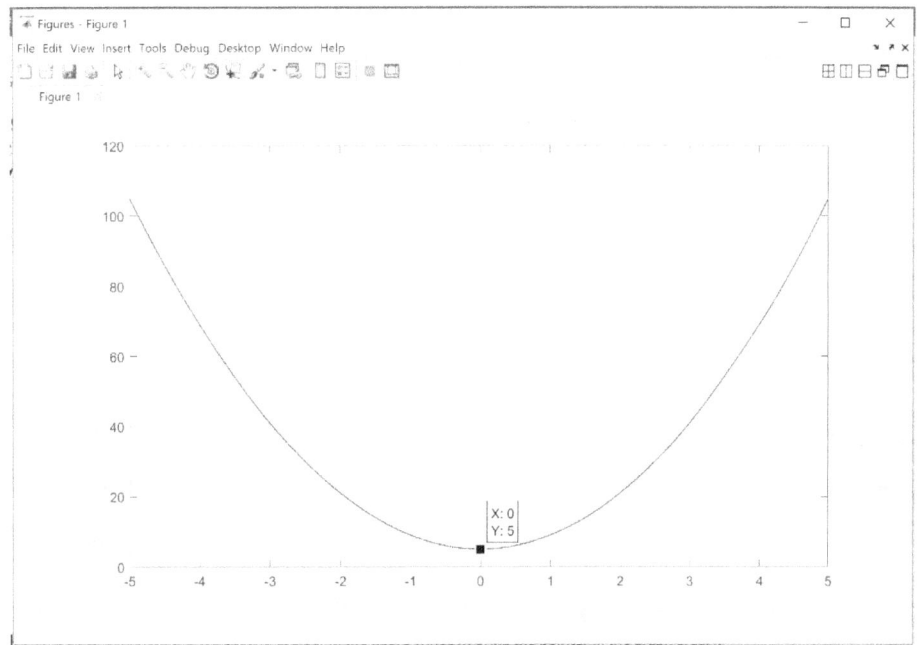

Left click on the **Data Cursor** in the top second toolbar in the figure window. Bring the pointer in the graph area, it converts into a '+' icon, hoover it over the plot. Bring the '+' to a point of interest on the plot and click the left button. This will put a solid square on the plot with its coordinates.

Standard and User Defined Functions
CALLING A STANDARD MATLAB FUNCTION

MATLAB has many standard functions which can be called (used) in any script file.

MATLAB m_file: _____

```
%appendix_M_5.m
x=0:9;
y1=x: y2=sqrt(x);      %call a standard MATLAB function to calculate the square root of a number
plot(x, y1, x, y2)
```

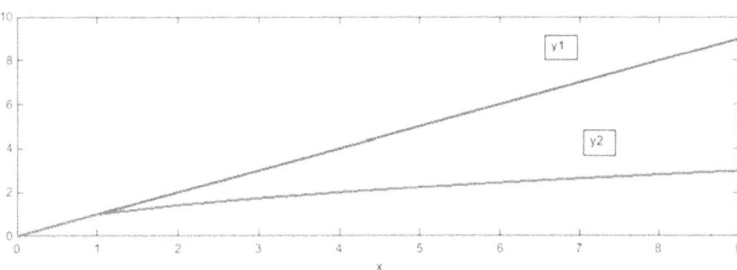

USER DEFINED FUNCTION

We will write a function "my_squarer()" function.
Type in the following file as a new function file and **save it in a file my_squarer.m**:

```
% to calculate the square of an array
function y= my_squarer(x)
        y= x.*x;
end
```

Do not Run the above function file.

Then write the following script file to call the above function.
MATLAB m_file: _____

```
%EET06_act3_4.m
clear                    %clear all previous variables in the workspace
x1=9.0;
y1=my_squarer(x1)        %call the function
x2=0:10;                 %x2 is an array
y2=my_squarer(x2);       %call the function
plot(x2, y2, 'linewidth', 1)
```

```
y1 =
    81
```

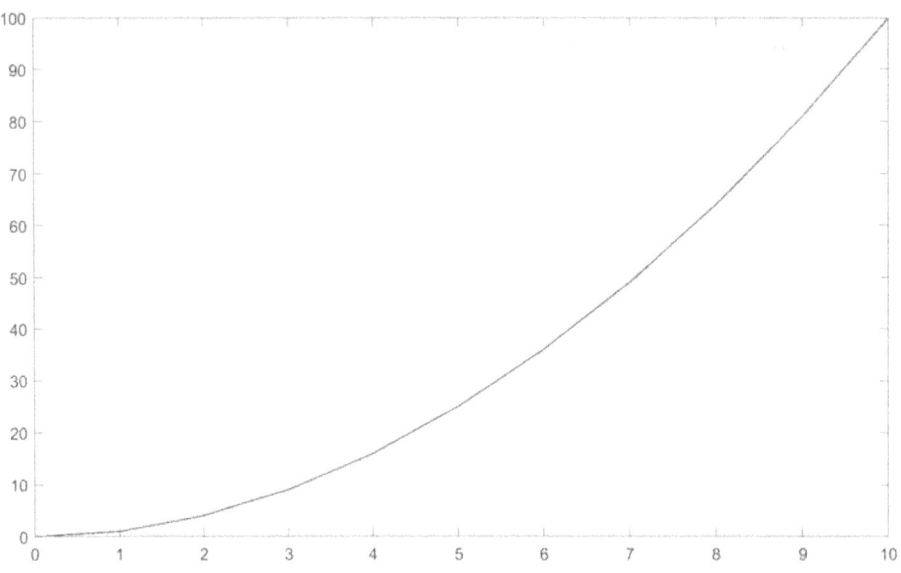

Appendix B

COMPLEX MATH

Note: Most of the material in this lab is adapted from the MATLAB help documentation.

Equipment/Software needed per station
-    MATLAB R2012 or higher

Theory
A complex number consists of two parts: the real part and the imaginary part:

$$C = a + jb$$
In MATLAB   C = a +1j * b

Function	Result	MATLAB Std. Function
Real part:	$= a$	$real(C)$
Imaginary part:	$= b$	$imag(C)$
Magnitude:	$= \sqrt{a^2 + b^2}$	$abs(C)$
Phase angle:	$= tan^{-1}\left[\frac{b}{a}\right]$	$angle(C)$

j- math : $j = \sqrt{-1}$ $j^2 = -1$ $j^3 = -j$ $\frac{1}{j} = -j$

Complex Conjugate:

$$d = a + jb \qquad d4 = conj(d) = \cos a - j \sin b$$

Adding/Subtracting two complex numbers:

$$d1 = a + jb \qquad d2 = p + jq \qquad ds = d1 + d2 = (a + p) + j(b + q)$$

Multiplying/Dividing two complex numbers:

$$d1 = a + jb \qquad d2 = p + jq \qquad ds = d1 \cdot d2 = (ap - bq) + j(aq + bp)$$

Complex Exponent:

$$d = a + jb \qquad d4 = e^d = e^a \cdot e^{jb} = e^a \cdot (\cos b + j \sin b)$$

Complex Math Functions in MATLAB:

abs( )	Absolute value and complex magnitude
angle( )	Phase angle
complex( )	Construct complex data from real and imaginary components
conj( )	Complex conjugate

i	Imaginary unit
imag( )	Imaginary part of complex number
isreal( )	Check if input is real array
real( )	Real part of complex number

Find the details of all the above functions by typing in the Command window:

>> help abs

MATLAB worksheet_example1_____

Enter the following statements in the command window and then press *Enter*. The MATLAB returns the result, shown in highlight. Try them and show the codes and your results.

```
>> c = 3 + 1j * 4
c =   3.0000 + 4.0000i


>> abs(c)
ans =   5


>> angle(c)
ans =   0.9273


>> angle(c)     %angle in radians
ans =   0.9273


>> real_part=5;     %terminating a statement by ; executes but does not show the result
>> imag_part=6;
>> c1= complex(real_part, imag_part)
c1 =  5.0000 + 6.0000i
>> conj(c1)
ans =  5.0000 - 6.0000i


>> c2 = 1/c1        % reciprocal of c1, note the reciprocal has the real number of opposite polarity
c2 =  0.0820 - 0.0984i


>> d1=3+1j*4, d2=5+1j*6     %Note that we can write two statements on a line, separated by ','
d1 =  3.0000 + 4.0000i
d2 =  5.0000 + 6.0000i


>> w1 = d1+d2
w1 =  8.0000 +10.0000i


>> w2=d1-d2
w2 = -2.0000 - 2.0000i


>> w3=d1*d2
w3 = -9.0000 +38.0000i


>> w4=d1/d2
```

w4 =   0.6393 + 0.0328i

>>d = 3+1j*4
d =   3.0000 + 4.0000i

>> d4=exp(d)   %exp is a built-in exponential function
d4 = -13.1288 -15.2008i

>> d5=exp(real(d))*(cos(imag(d))+1j*sin(imag(d)))   %verification
d5 = -13.1288 -15.2008i

Task # 1

A circuit is shown below:

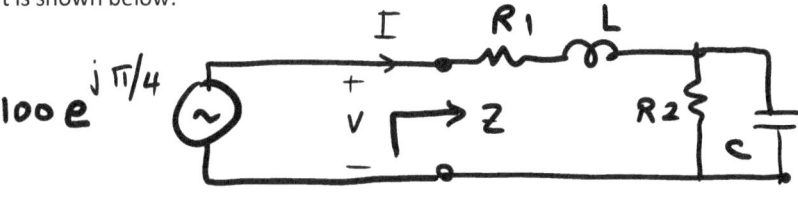

$$Z1 = R1 + j2\pi fL \qquad R1 = 2 \text{ ohms} \qquad L = 1 \text{ mH} \qquad f = 60 \text{ Hz}$$

$$Z2 = \frac{R2}{\frac{1}{j2\pi fC}} \qquad R2 = 4 \text{ ohms} \qquad C = 0.1 \text{ μF}$$

Determine the following using MATLAB and then verify the results by hand calculation (or use calculator). Submit the code and results.
a)   The expressions for Z, V and I
b)   Magnitude of Z
c)   Phase angle of Z
d)   Find the amplitude and phase of current I.

Matrix Math

In the MATLAB environment, a matrix is a rectangular array of numbers. Special meaning is sometimes attached to 1-by-1 matrices, which are scalars, and to matrices with only one row or column, which are vectors. MATLAB has other ways of storing both numeric and nonnumeric databut in the beginning, it is usually best to think of everything as a matrix.

Entering Matrices (ref: MATLAB Help)
The best way for you to get started with MATLAB is to learn how to handle matrices. Start MATLAB and **follow along** with each example (Submit codes and your results).

You can enter matrices into MATLAB in several different ways:
> Enter an explicit list of elements.
> Load matrices from external data files.
> Generate matrices using built-in functions.

Create matrices with your own functions and save them in files.

Start by entering Dürer's matrix as a list of its elements. You only have to follow a few basic conventions:
Separate the elements of a row with blanks or commas.
Use a semicolon, ; , to indicate the end of each row.
Surround the entire list of elements with square brackets, [ ].

To enter Dürer's matrix, simply type in the Command Window

```
>>A = [16, 3, 2, 13; 5, 10, 11, 8; 9, 6, 7, 12; 4, 15, 14, 1]
A =
   16    3    2   13
    5   10   11    8
    9    6    7   12
    4   15   14    1
```

Once you have entered the matrix, it is automatically remembered in the MATLAB workspace. You can refer to it simply as A. Now that you have A in the workspace, take a look at what makes it so interesting. Why is it magic?

sum, transpose, and diag
You can take the sum along any row or column, or along either of the two main diagonals. Let us do that using MATLAB. The first statement to try is

```
>>sum(A)                 %sum of elements in each column
ans =   34   34   34   34
```

When you do not specify an output variable, MATLAB uses the variable *ans*, short for answer, to store the results of a calculation. You have computed a row vector containing the sums of the columns of A.

How about the row sums? MATLAB has a preference for working with the columns of a matrix, so one way to get the row sums is to transpose the matrix, compute the column sums of the transpose, and then transpose the result. For an additional way that avoids the double transpose use the dimension argument for the sum function.

MATLAB has two transpose operators.
The apostrophe operator (e.g., A') performs a complex conjugate transposition. It flips a matrix about its main diagonal, and also changes the sign of the imaginary component of any complex elements of the matrix.
The dot-apostrophe operator (e.g., A.'), transposes without affecting the sign of complex elements. For matrices containing all real elements, the two operators return the same result.

```
>>A'
ans =
   16    5    9    4
    3   10    6   15
    2   11    7   14
   13    8   12    1

>>sum(A')'
ans =
   34
   34
```

34
34

The sum of the elements on the main diagonal is obtained with the sum and the diag functions:

>>diag(A)
ans =
  16
  10
   7
   1

>>sum(diag(A))
ans =
  34

A function originally intended for use in graphics, fliplr, flips a matrix from left to right: Try it.

>>fliplr(A)

The following sections continue to use this matrix to illustrate additional MATLAB capabilities.

Subscripts

The element in row i and column j of A is denoted by A(i, j). For example, A(4, 2) is the number in the fourth row and second column. For the magic square, A(4, 2) is 15. So to compute the sum of the elements in the fourth column of A, type

>>A(1, 4) + A(2, 4) + A(3, 4) + A(4, 4)
ans =
   34
But it is not the most elegant way of summing a single column.

If you try to use the value of an element outside of the matrix, it is an error:

>>t = A(4, 5)
Index exceeds matrix dimensions.

Conversely, if you store a value in an element outside of the matrix, the size increases to accommodate the newcomer:

>>X = A;
>>X(4,5) = 17
X =
  16   3   2 13   0
   5 10 11   8   0
   9   6   7 12   0
   4 15 14   1 17

The Colon Operator

The colon ':' is one of the most important MATLAB operators. It occurs in several different forms. The expression

```
>>1:10    %is a row vector containing the integers from 1 to 10:
1  2  3  4  5  6  7  8  9  10
```

To obtain non-unit spacing, specify an increment. For example,

```
>>100:-7:50
100  93  86  79  72  65  58  51
```

```
>>0:pi/4:pi
0   0.7854   1.5708   2.3562   3.1416
```

Subscript expressions involving colons refer to portions of a matrix:

```
>>A(1:k, j);                %is the first k elements of the jth column of A. Thus:
>>sum(A(1:4, 4)); %computes the sum of the fourth column.
```

However, there is a better way to perform this computation. The colon by itself refers to all the elements in a row or column of a matrix and the keyword end refers to the last row or column. Thus:

```
>>sum(A(:, end))  %computes the sum of the elements in the last column of A:
ans =
   34
```

Generating Matrices

MATLAB provides four functions that generate basic matrices.

zeros( )	All zeros
ones( )	All ones
rand( )	Uniformly distributed random elements
randn( )	Normally distributed random elements

Here are some examples:

```
>>Z = zeros(2, 4)
Z =
   0  0  0  0
   0  0  0  0
```

```
>>F = 5*ones(3, 3)
F =
   5  5  5
   5  5  5
   5  5  5
```

```
>>N = fix(10*rand(1, 10))
N =
   9  2  6  4  8  7  4  0  8  4
```

```
>>R = randn(4, 4)
R =
```

```
 0.6353   0.0860  -0.3210  -1.2316
-0.6014  -2.0046   1.2366   1.0556
 0.5512  -0.4931  -0.6313  -0.1132
-1.0998   0.4620  -2.3252   0.3792
```

The load Function

The load function reads binary files containing matrices generated by earlier MATLAB sessions, or reads text files containing numeric data. The text file should be organized as a rectangular table of numbers, separated by blanks, with one row per line, and an equal number of elements in each row. For example, outside of MATLAB, create a text file containing these four lines:

```
16.0    3.0    2.0   13.0
 5.0   10.0   11.0    8.0
 9.0    6.0    7.0   12.0
 4.0   15.0   14.0    1.0
```

Save the file as magik.dat in the current directory. The statement

>>load magik.dat

reads the file and creates a variable, magik, containing the example matrix.

An easy way to read data into MATLAB from many text or binary formats is to use the Import Wizard.

Saving Code to a File

You can create matrices using text files containing MATLAB code. Use the MATLAB Editor or another text editor to create a file containing the same statements you would type at the MATLAB command line. Save the file under a name that ends in .m.

For example, create a file in the current directory named magik.m containing these five lines:

```
A = [16.0    3.0    2.0   13.0
      5.0   10.0   11.0    8.0
      9.0    6.0    7.0   12.0
      4.0   15.0   14.0    1.0 ];
```

The statement
>>magik
reads the file and creates a variable, A, containing the example matrix.

Concatenation

Concatenation is the process of joining small matrices to make bigger ones. In fact, you made your first matrix by concatenating its individual elements. The pair of square brackets, [], is the concatenation operator. For an example, start with the 4-by-4 magic square, A, and form

>>B = [A, A+32; A+48, A+16]   %The result is an 8-by-8 matrix, obtained by joining the
                          four submatrices:

```
B =
  16    3    2   13   48   35   34   45
   5   10   11    8   37   42   43   40
   9    6    7   12   41   38   39   44
   4   15   14    1   36   47   46   33
```

```
64  51  50  61  32  19  18  29
53  58  59  56  21  26  27  24
57  54  55  60  25  22  23  28
52  63  62  49  20  31  30  17
```

Deleting Rows and Columns

You can delete rows and columns from a matrix using just a pair of square brackets. Start with

```
>>X = A;
>>X(:,2) = []   % to delete the second column of X
X =
  16   2  13
   5  11   8
   9   7  12
   4  14   1
```

If you delete a single element from a matrix, the result is not a matrix anymore. So, expressions like

```
>>X(1,2) = []
```

result in an error. However, using a single subscript deletes a single element, or sequence of elements, and reshapes the remaining elements into a row vector. So

```
>>X(2:2:10) = []
X =
  16   9   2   7  13  12   1
```

Linear Algebra

Informally, the terms matrix and array are often used interchangeably. More precisely, a matrix is a two-dimensional numeric array that represents a linear transformation. The mathematical operations defined on matrices are the subject of linear algebra.

Adding a matrix to its transpose produces a symmetric matrix:

```
>>A + A'
ans =
  32   8  11  17
   8  20  17  23
  11  17  14  26
  17  23  26   2
```

The multiplication symbol, *, denotes the matrix multiplication involving inner products between rows and columns. Multiplying the transpose of a matrix by the original matrix also produces a symmetric matrix:

```
>>A'*A
ans =
  378  212  206  360
  212  370  368  206
  206  368  370  212
  360  206  212  378
```

The determinant of this particular matrix happens to be zero, indicating that the matrix is singular:

>>d = det(A)
d =
   0

Because the matrix is singular, it does not have an inverse. If you try to compute the inverse with
>>X = inv(A)
Warning: Matrix is close to singular or badly scaled.
     Results may be inaccurate. RCOND = 9.796086e-018.

Arrays
When they are taken away from the world of linear algebra, matrices become two-dimensional numeric arrays. Arithmetic operations on arrays are done element by element. This means that addition and subtraction are the same for arrays and matrices, but that multiplicative operations are different. MATLAB uses a dot, or decimal point, as part of the notation for multiplicative array operations.

The list of operators includes
+   Addition
-          Subtraction
.*   Element-by-element multiplication
./   Element-by-element division
.\   Element-by-element left division
.^   Element-by-element power
.'   Unconjugated array transpose

If the Dürer magic square is multiplied by itself with array multiplication

>>A.*A
ans =
  256   9   4 169
   25 100 121   64
   81   36   49 144
   16 225 196   1

Building Tables
Array operations are useful for building tables. Suppose n is the column vector
The elementary math functions operate on arrays element by element. So format short g

>>x = (1:0.1:2)';
>>logs = [x, log10(x)]   %builds a table of logarithms.
 logs =
    1.0   0
    1.1   0.04139
    1.2   0.07918
    1.3   0.11394
    1.4   0.14613
    1.5   0.17609
    1.6   0.20412

1.7	0.23045
1.8	0.25527
1.9	0.27875
2.0	0.30103

Appendix C

OBJECT ORIENTED PROGRAMMING IN MATLAB: PRIMER

TOOL:
>	MATLAB 2017

PRE-REQUISITES

Background in simple programming using MATLAB is required.

WHY OBJECT ORIENTED PROGRAMMING IN MATLAB/SIMULINK

>	**Objects_lec05.pdf**

CREATE A SIMPLE MATLAB OBJECT

```
Syntax:
classdef BasicClass
  properties
    Value
  end
  properties
    V
  end
  methods
    function r = roundOff(obj)
      r = round([obj.Value],2);
    end
    function r = multiplyBy(obj,n)
      r = [obj.Value] * n;
    end
  end
  methods
    function r = exponent(obj, n)
        r = [obj.Value] .^ n;
      end
  end
end
```

```
classdef ASimpleObject
  properties
    Value
```

```matlab
  end
    methods
    function obj = ASimpleObject(val)
      if nargin > 0
       if isnumeric(val)
         obj.Value = val;
       else
         error('Value must be numeric')
       end
      end
    end
  end
  methods
    function r = roundOff(obj)
      r = round([obj.Value],2);
    end
    function r = multiplyBy(obj,n)
      r = [obj.Value] .* n;
    end
    function r = exponent(obj, n)
      r = [obj.Value] .^ n;
    end
  end
end
```

Object Programming in MATLAB in the command window:

```matlab
>> a=ASimpleObject(4)
a =
  ASimpleObject with properties:
    Value: 4

>> a.Value
ans =    4
>> b=roundOff(a)
b =    4

>> a=ASimpleObject(4.234)
a =   ASimpleObject with properties:
    Value: 4.2340

>> a.Value
ans =
    4.2340

>> b=roundOff(a)
b =    4.2300

>> c=exponent(a, 3)
```

c =   75.9019

OBJECT TO OPERATE ON ARRAYS

>> u=0:.01:5.0;
>>a=ASimpleObject(u)
Check all methods

OBJECT TO OPERATE ON RANDOM ARRAYS

>> u=0+5*rand(10,1)   % generate 10 random values between 0 to 5
>>a=ASimpleObject(u)
Check all methods

ASSIGNMENT

1)   modify the ASimpleObject, test for a single value (4.8746)
- a.   to roundoff up to 3 spaces.
- b.   Add a method to find squareroot of the value
- c.   Add a function to divide a value by another number

2) Create an object, *MyFirstObject*, to operate on a set of 5 values (1-2 digit numbers), that will
- a) calculate the sum of all values
- b) calculate the mean value and
- c) reorder the set of values in the descending order, top being the highest value.

Appendix D

DOWNLOADING AND GETTING STARTED

Important

Author presumes that readers have a working knowledge of programming in MATLAB, if not please go to Appendix A and do some practice.

Save all Phasor Tool Book functions, see Appendix E, in a folder of your convenience. You can also download all Phasor Tool Box functions from https://professorjaiagrawal.weebly.com/ ptb_functions.docx.

You also need to download the arrow function from the website (free download), https://www.mathworks.com/matlabcentral/fileexchange/?utf8=%E2%9C%93&term=arrow

Add path in your all MATLAB codes in this book. Example is as given below where my folder is jp_MATLAB :

___

GETTING STARTED

We will start with a simple example of ac circuit calculation using livescript in MATLAB using the Phasor Tool Book.

MATLAB_Ex_9
An ac electric circuit is shown below. Find and plot the phasor for voltage V_1 :

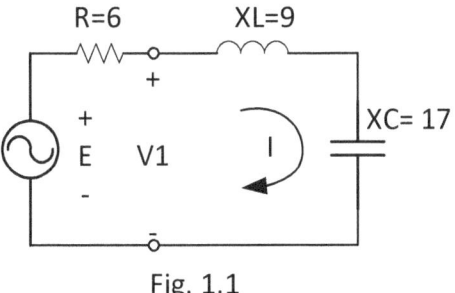

Fig. 1.1

___

Line 1   %PTB2_Ex_0.mlx
Line 2   addpath('c:\Users\jpagrawa\Documents\Aschool\jp_MATLAB_Circuits')
Line 3   f=1; T=1/f;
Line 4   clf, clear
Line 5   ZR=phasor(6, 0);
Line 6   ZL=phasor(9, 90);
Line 7   ZC=phasor(17, -90);
Line 8   ZT=ZR+ ZL+ ZC
ZT =

phasor with properties:

   Mag: 10.0000
  phase: -53.1301
Line 9   E=phasor(50, 30);
Line 10 ZLC=ZL + ZC
ZLC =
  phasor with properties:

   Mag: 8
  phase: -90
Line 11 V1=E * ZLC / ZT
V1 =
  phasor with properties:

   Mag: 40.0000
  phase: -6.8699
Line 12 phplot(V1)
Current plot held
Current plot released

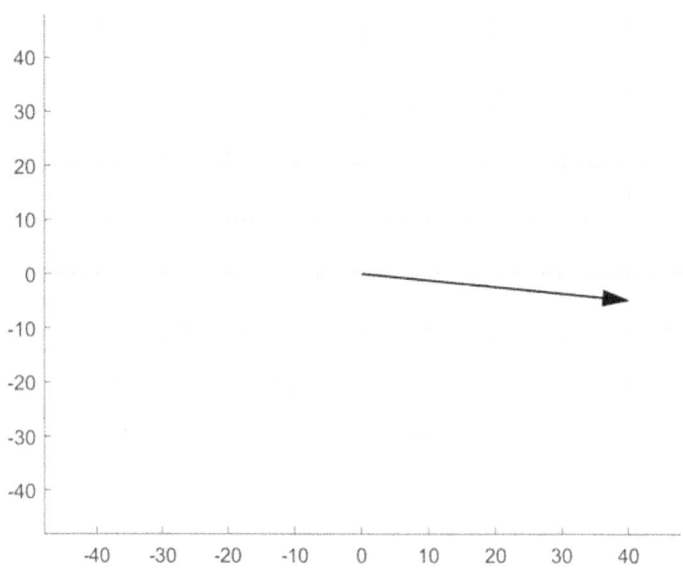

The above example can also be executed as .m file in MATLAB.

PROGRAM EXPLANATION

Line 1   Comment line for your reference where you have stored the current script (MATLAB code).
Line 2   The path is where you have stored all Phasor Tool Box functions for future use. It is advisable to write as many comments as needed in order to understand the logic of the code, say after a month. The results of executing the .mlx scripts are shown without line numbers. The readers are advised to code the above program, run and verify.

All examples in this book assume that you would automatically have done this on your computer any time you use them, therefore, this line will be omitted in the MATLAB codes hereafter. Alternatively save the Phasor Tool Box function in your current working folder and do all future exercises in this folder.

Line 3   f is the operating frequency and T is the period of the operating frequency.

Line 4   **clf** to clear all previous graphs and **clear** is to delete all previous variables in the MATLAB workspace.

Line 5   **ZR** is defined as a polar impedance (resistive) with 6 ohms and 0 phase, an object under the class phasor.

Line 6   **ZL** is defined as a polar impedance (inductive) with 9 ohms and 90 deg phase, an object under the class phasor.

Line 7   **ZC** is defined as a polar impedance (capacitive) with 17 ohms and -90 deg phase, an object under the class phasor.

Line 8   **ZT** is the total impedance as seen by the source E. Since there is no ; at the end of the line, the line is executed and the result is shown directly below the line 8. ZT is a polar impedance with magnitude of 10 ohms with a phase of -53.1301 deg.

Line 9   **E** is defined as a phasor voltage source with 50 V rms and 30 deg phase, an object under the class phasor.

Line 10   **ZLC** is the sum of ZL and ZC impedances. Since there is no ; at the end of the line, the line is executed and the result is shown directly below the line 10. ZLC is a polar impedance with magnitude of 8 ohms with a phase of -- 90 deg.

Line 11   **V1** is the phasor voltage across the series combination of sum of ZL and ZC impedances, using the voltage divider rule. Since there is no ; at the end of the line, the line is executed and the result is shown directly below the line 11. V1 is a phasor voltage with magnitude of 40 Vrms and a phase of -6.8699 deg.

Line 12   This code gives a plot of phasor **V1** on the complex plane. The **V1** phasor is an arrow beginning at the the origin (0, 0) point, at an angle of 6.8699 in the clockwise direction from the horizontal line and a length equal to the magnitude of phasor **V1**.

Appendix E

PHASOR TOOL BOOK FUNCTIONS

```
function VT=add_graph(V1, varargin)
%{
Phasor V1 is plotted on the complex plane from the origin (0, 0), then V2 is added to the end of V1. Phasor V3 is
added to the end of V2, and so on. Finally, the phasor VT, the algebraic sum of all V1, V2, ..., is drawn from the origin
to the end of the last vector  added. No limit on the number of vectors for addition. VT is in red color.
%written by Prof. Jai Agrawal
%}
  clf
  N=nargin;
  V(1)=V1;
  RT(1)=(V(1).Mag)*cosd(V(1).phase);
  IT(1)=(V(1).Mag)*sind(V(1).phase);
  S1=V(1).Mag;
  m=1.2*S1;
  axis([-m,m,-m, m]);
  grid;
  arrow([0, 0], [RT(1), IT(1)], 'color', 'black', 'linewidth', 1);
  for i=2:N
          V(i)=varargin{i-1};
          RT(i)=RT(i-1)+(V(i).Mag)*cosd(V(i).phase);
          IT(i)=IT(i-1)+(V(i).Mag)*sind(V(i).phase);
          m=max([m, sqrt(RT(i)^2+IT(i)^2)]);
          axis([-m,m,-m, m])
          arrow([RT(i-1), IT(i-1)], [RT(i), IT(i)], 'color', 'blue', 'linewidth', 1)
  end
  m=1.2*m;
  axis([-m,m,-m, m])
  arrow([0, 0], [RT(N), IT(N)], 'color', 'red', 'linewidth', 1)
  axis('square');
  VT=phasor(RT(N), IT(N), 'x2po');
end
```

```
function [h,yy,zz] = arrow(varargin)
```

```
function C = coeff(c)
%coefficient c is used for multiplying the magnitude of a phasor
%written by Prof. Jai P Agrawal
C=phasor(c, 0);
end
```

```
function [ Z1, Z2, Z3 ] = delta2wye( ZA, ZB, ZC, varargin )
```

```
%Delta-connected impedances to Y-connected impedances
%[ Z1, Z2, Z3 ] = delta2wye( ZA, ZB, ZC )      input arguments are all polar impedances
%[ Z1, Z2, Z3 ] = delta2wye( ZA, ZB, ZC, 'cx' )      input arguments are all complex impedances
%output impedances are all in polar format
% written by Prof. Jai Agrawal
 if nargin==3
        type=' ';
     else
        type=varargin{1};
 end
div=ZA+ZB+ZC;
Z1=ZB*ZC/div;
Z2=ZA*ZC/div;
Z3=ZA*ZB/div;
if(type=='cx')
   Z1=x2ph(Z1, 'po');
   Z2=x2ph(Z2, 'po');
   Z3=x2ph(Z3, 'po');
end
end
```

```
function [ EAN, EBN, ECN] = line2phase( EAB, EBC, ECA)
%written by Prof. Jai Agrawal
%EAB, EBC and ECA are 3-phase line voltage phasors
%EAN, EBN, ECN are the phase voltage phasors
c=coeff(sqrt(3));
p=rotate(30);
EAN=p*EAB/c ;
EBN=p*EBC/c;
ECN=p*ECA/c;
end
```

```
function  ZT = parallelZ(Z, varargin )
%{
written by Prof. Jai P Agrawal
Finds parallel combination of two or more impedances or admittances in polar or complex forms. Z is an array
of either impedances (Z's) or admittances (Y's). Output of this function ZT is a polar vector for impedance
or admittance.
ZT=parallelZ(Z)        to find parallel combination of all elements in an array of impedances, in polar form.
ZT=parallelZ(Z, 'cx')   to find parallel combination of all elements in an array of impedances, in complex form
%}
N=length(Z);
 if nargin==1
        type=' ';
     else
        type=varargin{1};
 end
  if(type==' ')
```

```
  YT=coeff(0);
  one=coeff(1);
  for i=1:N
     YT=YT+one/Z(i);
  end
  ZT=one/YT;
end

if(type=='cx')
  YT=0;
  for i=1:N
     YT=YT+1./Z(i);
  end
  Z=1./YT;
  ZT=phasor(real(Z), imag(Z), 'x2po');
end
end
```

```
function Y = ph2pu(A, Sb, Vb, type)
%A is a phasor, Sb and Vb are scalars
%Y=ph2pu(A, Sb, Vb, 'V')    A is a voltage phasor
%Y=ph2pu(A, Sb, Vb, 'I')    A is a current phasor
%Y=ph2pu(A, Sb, Vb, 'Z')    A is an impedance polar
%Y=ph2pu(A, Sb, Vb, 'S')    A is the apparent power phasor

Ib=Sb/Vb;         %base current
Zb=Vb ^2 /Sb;     %base impedance
c1=coeff(Vb);  c2=coeff(Ib);
c3=coeff(Zb);  c4=coeff(Sb);
Y=coeff(0);
if(type=='V')
Y=A / c1;         %per unit voltage
end
if(type=='I')
Y=A / c2;         %per unit voltage
end
if(type=='Z')
Y=A / c3;         %per unit voltage
end
if(type=='S')
Y=A /c4  ;        %per unit voltage
end
end
```

```
function x = ph2x(obj, varargin)
%x=ph2x(obj)         Convert a phasor matrix to complex form
%x=ph2x(obj, 'po')   Convert a polar matrix to complex form
%written by Prof. Jai Agrawal
```

```
  if nargin==1
      type=' ';
    else
      type=varargin{1};
    end
        [M, N]=size(obj);
      [mag, th]=prop(obj);
       mag=[mag.*sqrt(2)];
        if(type == 'po')
          mag=[mag./sqrt(2)];
        end
        x=mag.*cosd(th)+j*mag.*sind(th);
    end
```

```
function ZTp = inputZ(Z, V, varargin)
%{
```
Finds the input impedance between two terminals A-A' in a circuit. Z is the mesh impedance matrix obtained from setting all voltages sources to zero and 'opening' the current sources. Remove the component at A-A' terminals and connect an external voltage source Vx=1.0 ?0 across it. The positive terminal of Vx is connected to the terminal A. The V is a Mx1 vector containing nothing but Vx at the appropriate place. Output ZTp is a polar value. This function can also be used to find the output impedance as seen between terminals A-A'.
```
%}
%ZTp = inputZ(Z, V) Z and V are polar matrix and phasor respectively.
%ZTp = inputZ(Z, V, 'cx')   Z and V are complex matrix and complex vector respectively.

  if nargin==2
      type=' ';
    else
      type=varargin{1};
  end

[M, N]=size(Z);
m1=0; m2=0;
if (type=='cx')
  IT=Z\V;
  for i=1:M
    if V(i) >0
       m1=i;
    elseif V(i)<0
       m2=i;
    end
  end
  if m1>0 & m2==0
    ZT=V(m1)/IT(m1);
  end
  if m1==0 & m2>0
    ZT=V(m2)/IT(m2);
  end
  if m1>0 & m2>0
```

```
      ZT=V(m1)/(IT(m1)-IT(m2));
    end
  ZTp=x2ph(ZT, 'po');
end


 if (type=='  ')
    IT=Z\V;
    for i=1:M
      if  V(i).phase==0  & V(i).Mag>0
         m1=i;
      elseif  V(i).phase==180 & V(i).Mag >0
         m2=i;
      end
    end
  if m1>0 & m2==0
    ZT=V(m1)/IT(m1);
  end
  if m1==0 & m2>0
    ZT=V(m2)/IT(m2);
  end
  if m1>0 & m2>0
    ZT=V(m1)/(IT(m1)-IT(m2));
  end
  ZTp=ZT;
 end
end
```

```
function [ EAB, EBC, ECA] = phase2line(EAN, EBN, ECN)
%written by Prof. Jai Agrawal
%EAB, EBC and ECA are 3-phase line voltage phasors
%EAN, EBN, ECN are the phase voltage phasors
    EAB=EAN - EBN;
    EBC=EBN - ECN;
    ECA=ECN - EAN;
 end
```

```
classdef phasor
  properties
    Mag
    phase
  end
    methods
     function obj = phasor(val1, val2, varargin)
      if nargin > 0
        if isnumeric(val1)
         obj.Mag = val1;
         obj.phase = val2;
```

```matlab
        if nargin==2
            type=' ';
        else
            type=varargin{1};
        end
        if ((type=='x2po') |( type=='x2ph'))
            x=val1+j*val2;
            obj.Mag=abs(x);
            %
            if (type=='x2ph')
                obj.Mag=obj.Mag/sqrt(2);
            end
            %
            obj.phase=atand(val2/val1);
            if(val1<0)
                obj.phase=180+obj.phase;
            end
        end

    else
        error('Value must be numeric')
    end
   end
  end
 end

methods
  function[ B, C] = prop(obj)
   [M, N]=size(obj);
   for i=1:M
     for j=1:N
        B(i, j)=obj(i, j).Mag;
        C(i, j)=obj(i, j).phase;
     end
   end
  end

  function Y= conj(X)
    Y=X;
    Y.phase=-X.phase;
   end
  function r = uminus(obj)
    rm= obj.Mag;
    rp=obj.phase+180;
    r=phasor(rm, rp);
   end
  function r = plus(o1,o2)
    r1=[o1.Mag]*cosd([o1.phase])+[o2.Mag]*cosd([o2.phase]);
    i2=[o1.Mag]*sind([o1.phase])+[o2.Mag]*sind([o2.phase]);
```

```
      rm= sqrt( r1^2 + i2^2);
      rp=atand(i2/r1);
      if(r1<0)
        rp=180+atand(i2/r1);
      end
      r=phasor(rm, rp);
    end
   function r = minus(o1,o2)
      r1=[o1.Mag]*cosd([o1.phase])-[o2.Mag]*cosd([o2.phase]);
      i2=[o1.Mag]*sind([o1.phase])-[o2.Mag]*sind([o2.phase]);
      rm= sqrt( r1^2 + i2^2);
      rp=atand(i2/r1);
      if(r1<0)
        rp=180+atand(i2/r1);
      end
      r=phasor(rm, rp);
    end
   function r = mtimes(o1,o2)
      rm = [o1.Mag]*[o2.Mag];
      rp=[o1.phase]+[o2.phase];
      r=phasor(rm, rp);
    end
   function r = mrdivide(o1,o2)
      rm = [o1.Mag] / [o2.Mag];
      rp=[o1.phase]-[o2.phase];
      r=phasor(rm, rp);
    end
   function r = mldivide(o1,o2)
      A=ph2x(o1, 'po');
      B=ph2x(o2, 'po');
      C=A\B;
      r=x2ph(C, 'po');
    end
  end
 end
end
```

```
function phplot(A, varargin)
%phplot(A)          plot of phasor matrix A in the complex plane
%phplot(A, 'cx')     plot of complex matrix A in the complex plane
%the second input argument is optional.
% written by Prof. Jai Agrawal

  if nargin==1
      type=' ';
    else
      type=varargin{1};
    end

[M, N]=size(A);
```

```
if(type=='cx')
 m=1.2*max(max(abs(A)));
 axis([-m, m, -m, m])
 axis('square')
 hold
  for i=1:M
   for j=1:N
     arrow([0,0], [real(A(i, j)), imag(A(i, j))], 'linewidth', 1)
   end
  end
 grid
 hold
else
   [mag, th]=prop(A);
   m=1.2*max(max(mag));
   axis([-m, m, -m, m])
   axis('square')
   hold
     for i=1:M
       for j=1:N
           arrow([0,0], [A(i, j).Mag*cosd(A(i, j).phase), A(i, j).Mag*sind(A(i, j).phase)], 'linewidth', 1)
       end
     end
     grid
     hold
     end
end
```

```
function [y, t]=phplot_signal( V, f, t1, t2 , varargin)
%[y, t]=phplot_signal( V, f, t1, t2)     plots the sinusoidal waveforms of Matrix of phasors V
%over a range of time between t1 and t2
%y and t are respectively the amplitude and the time matrices of the time domain signal of the phasor V
%y is the amplitudes of the matrix of phasors in V
%[y, t]=phplot_signal( V, f, t1, t2 , 'cx')     plots the sinusoidal waveforms of Matrix of phasors V
%over a range of time between t1 and t2
% written by Prof. Jai Agrawal
if nargin==4
      type=' ';
    else
      type=varargin{1};
end
[M, N]=size(V);
T=1/f;
if(type==' ')
[mag, thd]=prop(V);
  A=mag.*sqrt(2);
  th=deg2rad(thd);
end
```

```
if(type=='cx')
  A=abs(V);
  th=angle(V);
end
  m=1.2*max(max(A)) ;
  dT=(t2-t1)/200;
  t=t1:dT:t2;
  hold
  for i=1:M
    for j=1:N
      y=A(i, j).* sin(2*pi*f*t+th(i, j));
      plot(t, y, 'LineWidth', 1)
    end
  end
  hold
  grid
  axis([t1, t2, -m, m])
end
```

```
function Y = pu2ph(A, Sb, Vb, type)
%Converts per unit phasor A to the actual phasor/polar value. Sb is the rated apparent power,
%and Vb is the rated voltage. Both Sb and Vb are scalars. Y is the actual phasor/polar value.
%Y=pu2ph(A, Sb, Vb, 'V')      A is a voltage phasor
%Y=pu2ph (A, Sb, Vb, 'I')      A is a current phasor
%Y=pu2ph (A, Sb, Vb, 'Z')      A is an impedance polar
%Y=pu2ph (A, Sb, Vb, 'S')      A is the apparent power polar
%Written by Prof. Agrawal
Ib=Sb/Vb;          %base current
Zb=Vb^2/Sb;        %base impedance
c1=coeff(Vb);
c2=coeff(Ib);
c3=coeff(Zb);
c4=coeff(Sb);
Y=coeff(0);
if(type=='V')
Y=A * c1;          %per unit voltage
end
if(type=='I')
Y=A * c2 ;         %per unit voltage
end
if(type=='Z')
Y=A * c3 ;         %per unit voltage
end
if(type=='S')
Y=A * c4 ;         %per unit voltage
end
end
```

```matlab
function [Ss, Pp, Qq, Fp, ph]=PWR(V, I, varargin)
%{
Calculates the power crossing two terminals A-A' in an electrical circuit. V and I are respectively the rms voltage and
rms current phasors in the circuit at the terminals A-A'.  S is apparent power, Q is reactive power and P is real power
(not in phasor form), Fp is the power factor and ph indicates whether the power factor is lagging or leading type.
The type indicates the type of input arguments. All output variables are in polar format. It also displays a triangle plot
of S, P and Q polars.
[S, P, Q, Fp, ph] = PWR(V, I)          V and I are phasors
[S, P, Q, Fp, ph] = PWR(V, I, 'cx')    V and I are complex vectors
%written by Prof. Jai Agrawal
%}

  if nargin==2
      type=' ';
    else
      type=varargin{1};
  end

if(type==' ')
  S=V* conj(I);
  Sm=S.Mag; Sp=S.phase;
  P=Sm* cosd(Sp);
  Q=Sm* sind(Sp);
  triplot(S) ;
end
if(type=='cx')
  S=(1/2)*V*conj(I);
  P=real(S);
  Q=imag(S);
  Sm=abs(S);
  Sp=atand(Q/P);
  if(P<0)
       Sp=180+atand(P/Q);
   end
  triplot(x2ph(S, 'po'));
end

  Fp=P/Sm;
  ph1='lagging'; ph2='leading'; ph='unity';
   if (Q>0)
     ph=char(ph1);
     th=90;
   else
     ph=char(ph2);
     th=-90;
   end
   Q=abs(Q);

   Pp=phasor(P, 0);
```

```
    Qq=phasor(Q, th);
    Ss=phasor(Sm, Sp);

end
```

```
function [Ss, Pp, Qq, Fp, ph]=PWR_3line(VLL, IL)
%Calculates the powers in 3-phase circuits.
%Line to line voltages VLL and Line currents IL are phasor arrays of 3 elements for 3 phases.
%S is apparent power, Q is reactive power and P is real power and Fp is the power factor.
%The ph indicates whether the power factor is lagging or leading type. S, P and Q are all polar quantities.
%written by Prof. Jai Agrawal
  Vab=VLL(1); Vbc=VLL(2); Vca=VLL(3);
 Ia=IL(1); Ib=IL(2); Ic=IL(3);
  Sa=coeff(1/sqrt(3))*Vab *rotate(-30)*conj(Ia);
  Sb=coeff(1/sqrt(3))*Vbc *rotate(-30)* conj(Ib);
  Sc=coeff(1/sqrt(3))*Vca *rotate(-30)* conj(Ic);
  S=Sa+ Sb+ Sc;
  P=S.Mag * cosd(S.phase);
  Q=S.Mag * sind(S.phase);
  Sm=S.Mag;
  Sp=S.phase;
  triplot(S) ;
  Fp=P/Sm;
 ph1='lagging'; ph2='leading'; ph='unity';
  if (Q>0)
    ph=char(ph1);
    th=90;
   else
    ph=char(ph2);
    th=-90;
   end
  Q=abs(Q);
  Pp=phasor(P, 0);
  Qq=phasor(Q, th);
  Ss=phasor(Sm, Sp);
end
```

```
function [Ss, Pp, Qq, Fp, ph]=PWR_3phase(Vph, Iph)
%Calculates the powers in 3-phase circuits. Phase to neutral voltages Vph and phase currents Iph
%are phasor arrays of 3 elements for 3 phases. S is apparent power, Q is reactive power and P is
%real power and Fp is the power factor. The ph indicates whether the power factor is lagging or leading type.
%S, P and Q are all polar quantities.
%written by Prof. Jai Agrawal
  Van=Vph(1); Vbn=Vph(2); Vcn=Vph(3);
 Ian=Iph(1); Ibn=Iph(2); Icn=Iph(3);
  Sa=Van *conj(Ian);
  Sb=Vbn * conj(Ibn);
```

```matlab
  Sc=Vcn * conj(Icn);
  S=Sa+ Sb+ Sc;
  P=S.Mag * cosd(S.phase);
  Q=S.Mag * sind(S.phase);
  Sm=S.Mag;
  Sp=S.phase;
  triplot(S) ;
 Fp=P/Sm;
 ph1='lagging'; ph2='leading'; ph='unity';
  if (Q>0)
    ph=char(ph1);
    th=90;
   else
    ph=char(ph2);
    th=-90;
   end
   Q=abs(Q);
   Pp=phasor(P, 0);
   Qq=phasor(Q, th);
   Ss=phasor(Sm, Sp);
end
```

```matlab
function p = rotate(d)
%Rotate a phasor/polar by d degrees counterclockwise
%multiply a phasor by p.
%Written by Professor Jai Agrawal
p=phasor(1, d);
end
```

```matlab
function [ Eth,  Zth] = Thevenin( Z, E, Vx, type)
%Coded by Prof JP Agrawal
 %Finds the Thevenin equivalent circuit in an electrical network between terminals A-A'.
 %To achieve this, remove any component connected to terminals A-A', and instead connect
 %an external unity voltage source with zero phase Vx. = phasor(1, 0). Write the impedance matrix Z
 %with terminals with A-A' connected to Vx set to zero magnitude. E is source vector without Vx.
 %V is the source vector containing only external voltage Vx across A-A' terminal, Positive of Vx
 %connected to terminal A.
 %The output Eth is a voltage phasor and Zth is the polar impedance.
 %[Eth,  Zth] = Thevenin( Z, E, V, 'po')        Z is polar matrix, E and V are phasor vectors
 %[Eth,  Zth] = Thevenin( Z, E, V, 'cx')        Z, E and V are all in complex form
  [M, N]=size(Z);
  if (type=='cx')
    Zth=inputZ(Z, Vx, 'cx') ;               %Thevenin Impedance Zth, looking into a-a' terminal
    Im1=Z\E;                                %solution of Z Im=V,
    Ish=0;                                  %initialization, external source applied for calculating thevenin
    for i=1:M
      if Vx(i)<0
```

```
        Ish=Ish+Im1(i);              %short circuit current in branch a-a'
      elseif Vx(i)>0
        Ish=Ish-Im1(i);              %short circuit current in branch a-a'
     end
   end
   Ish=x2ph(Ish);
   Eth=Ish*Zth;
 end


 if (type=='po')
    Zth=inputZ(Z, Vx);               %Thevenin Impedance Zth, looking into a-a' terminal
    Im1=Z\E ;                        %solution of Z Im=V
    Ish=coeff(0);                    %initialization, external source applied for calculating thevenin

    for i=1:M
      if Vx(i).phase==180 & Vx(i).Mag>0
        Ish=Ish+Im1(i) ;             %short circuit current in branch a-a'
      elseif Vx(i).phase==0 & Vx(i).Mag>0
        Ish=Ish-Im1(i);             %short circuit current in branch a-a'
     end
    end
    Eth=Ish*Zth;                     %thevenin voltage Eth
  end
end
```

```
function triplot(A)
%A  is a phasor V or I, polar Z or Y, Apparent Power S %(Volt Ampere) in the polar form
%written by Prof. Agrawal
  R=(A.Mag)*cosd(A.phase);
  I=(A.Mag)*sind(A.phase);
  S=A.Mag;
  m=1.2*S;
axis([-m,m,-m, m])
grid
arrow([0, 0], [R, 0], 'color', 'black', 'linewidth', 1)
arrow([R, 0], [R, I], 'color', 'blue', 'linewidth', 1)
arrow([0,0], [R, I], 'color', 'red', 'linewidth', 1)
axis('square');
end
```

```
function [ ZA, ZB, ZC ] = wye2delta(Z1, Z2, Z3, varargin )
%Y-connected impedances to Delta-connected impedances
%[ ZA, ZB, ZC ] = wye2delta(Z1, Z2, Z3)          Y-connected polar
%impedances to Delta-connected polar impedances
%[ ZA, ZB, ZC ] = wye2delta(Z1, Z2, Z3, 'po')         Y-connected complex
%impedances to Delta-connected polar impedances
% written by Prof. Jai Agrawal
```

```matlab
  if nargin==3
      type=' ';
    else
      type=varargin{1};
  end

num=Z1*Z2+Z2*Z3+Z1*Z3;
ZA=num/Z1;
ZB=num/Z2;
ZC=num/Z3;
if(type=='cx')
  ZA=x2ph(ZA, 'po');
  ZB=x2ph(ZB, 'po');
  ZC=x2ph(ZC, 'po');
end
end
```

```matlab
function P = x2ph(x, varargin)
%written by Prof. Jai Agrawal
%P=x2ph(obj)            Convert a complex matrix to phasor matrix.
%P=x2ph(obj, 'po')      Convert a complex matrix to polar matrix.

    if nargin==1
      type=' ';
    else
      type=varargin{1};
    end
      [M, N]=size(x);
      rm=[abs(x)/sqrt(2)];
    if (type=='po')
      rm=rm*sqrt(2);
    end
        rp=[rad2deg(angle(x))];
        for j=1:M
        for i=1:N
          P(j, i)=phasor(rm(j, i), rp(j, i));
        end
        end
  end
```

```matlab
function [ Eth,  Zth] = Thevenin( Z, E, Vx, varargin)
%Coded by Prof JP Agrawal
  [M, N]=size(Z);
%Finds the Thevenin equivalent circuit in an electrical network between terminals A-A'.
%To achieve this, remove any component connected to terminals A-A', and instead connect
%an external unity voltage source with zero phase Vx. = phasor(1, 0). Write the impedance matrix Z
%with terminals with A-A' connected to Vx set to zero magnitude. E is source vector without Vx.
```

```
%V is the source vector containing only external voltage Vx across A-A' terminal, Positive of Vx
%connected to terminal A.
%The output Eth is a voltage phasor and Zth is the polar impedance.

%[Eth, Zth] = Thevenin( Z, E, Vx)    Z is polar matrix, E and V are phasor vectors
%[Eth, Zth] = Thevenin( Z, E, Vx, 'cx')   Z, E and V are all in complex form
if nargin==3
        type=' ';
    else
        type=varargin{1};
end
  if (type=='cx')
    Zth=inputZ(Z, Vx, 'cx') ;          %Thevenin Impedance Zth, looking into a-a' terminal
    Im1=Z\E;                           %solution of Z Im=V,
    Ish=0;                             %initialization, external source applied for calculating thevenin
    for i=1:M
      if Vx(i)<0
        Ish=Ish+Im1(i);                %short circuit current in branch a-a'
      elseif Vx(i)>0
        Ish=Ish-Im1(i);                %short circuit current in branch a-a'
      end
    end
    Ish=x2ph(Ish);
    Eth=Ish*Zth;
  end
  if (type==' ')
    Zth=inputZ(Z, Vx);                 %Thevenin Impedance Zth, looking into a-a' terminal
    Im1=Z\E ;                          %solution of Z Im=V
    Ish=coeff(0);                      %initialization, external source applied for calculating thevenin
    for i=1:M
      if Vx(i).phase==180 & Vx(i).Mag>0
        Ish=Ish+Im1(i) ;               %short circuit current in branch a-a'
      elseif Vx(i).phase==0 & Vx(i).Mag>0
        Ish=Ish-Im1(i);                %short circuit current in branch a-a'
      end
    end
    Eth=Ish*Zth;                       %thevenin voltage Eth
  end
end
```

```matlab
function [ ZA, ZB, ZC ] = wye2delta(Z1, Z2, Z3, type )
%Y-connected impedances to Delta-connected impedances
% written by Prof. Jai Agrawal
num=Z1*Z2+Z2*Z3+Z1*Z3;
ZA=num/Z1;
ZB=num/Z2;
ZC=num/Z3;
if(type=='cx')
    ZA=x2ph(ZA, 'po');
    ZB=x2ph(ZB, 'po');
    ZC=x2ph(ZC, 'po');
end
end
```

www.ingramcontent.com/pod-product-compliance
Lightning Source LLC
Chambersburg PA
CBHW081723220526
45468CB00008B/1950